"十四五"普通高等教育本科部委级规划教材

服装英语

（第4版）

马凯　郭平建　主编
吕逸华　审校

FUZHUANG YINGYU

中国纺织出版社有限公司

内 容 提 要

本书课文选自经典的英语原版书籍，融合更新更实用的服装理念，内容包括时装业演变、设计师工作、服装材料、纸样制作、服装生产、服装零售、服装广告、服装业职位、消费者需求与服装营销等，并辅以近期优秀的学术期刊文献节选作为补充阅读资料。涉及服装专业英语词类广、选材新颖、内容丰富、实用性强。通过学习本书，读者可以更快更轻松地掌握服装设计、生产和贸易的专业英语。

本书可作为高等院校服装设计、服装工程、服装史论、服装文化和服装贸易专业的英语教材，也可供广大服装工作者和业余爱好者阅读。

图书在版编目（CIP）数据

服装英语：中英文/马凯，郭平建主编. --4版. --北京：中国纺织出版社有限公司，2021.9
"十四五"普通高等教育本科部委级规划教材
ISBN 978-7-5180-8711-2

Ⅰ.①服… Ⅱ.①马…②郭… Ⅲ.①服装工业-英语-高等学校-教材-汉、英 Ⅳ.①TS941

中国版本图书馆CIP数据核字（2021）第141618号

责任编辑：李春奕　籍　博　　责任校对：王蕙莹
责任印制：王艳丽

中国纺织出版社有限公司出版发行
地址：北京市朝阳区百子湾东里A407号楼　邮政编码：100124
销售电话：010—67004422　传真：010—87155801
http://www.c-textilep.com
中国纺织出版社天猫旗舰店
官方微博 http://weibo.com/2119887771
三河市宏盛印务有限公司印刷　各地新华书店经销
1997年11月第1版　2000年6月第2版
2007年8月第3版　2021年9月第4版第1次印刷
开本：787×1092　1/16　印张：11.5
字数：187千字　定价：49.80元

凡购本书，如有缺页、倒页、脱页，由本社图书营销中心调换

前言 Preface

《服装英语》于 1997 年 11 月由中国纺织出版社出版，2000 年 6 月推出第 2 版，2007 年 8 月推出第 3 版，目前已第 18 次印刷。该教材问世以来，受到了服装专业广大师生的一致好评，在广大读者中产生了深远的影响，对培养服装专业人才的专业英语阅读能力起到了积极的作用。但是近些年来，随着服装产业数字化、网络化和智能化水平的不断提升，3D 打印、智能面料、智能芯片等新科技都开始在服装设计、生产与制造等领域大展拳脚，相关的新名词也大量涌入。因此，为了提高服装专业师生和广大服装工作者阅读服装英语文献、书刊的能力，与科技发展新时代接轨，我们在保留原有经典课文的基础之上，又补充了多篇与服装新科技相关的阅读资料，并更新了有关服装生产和外贸的实用文体，编写出该书的第 4 版。

全书共 10 章，每章包括两个部分，可作为服装设计、服装工程、服装史论、服装文化和服装贸易专业的英语教材。第 1 部分收录专业技术文章，用来培养读者阅读和翻译服装专业英语资料的能力。内容涉及服装史、服装设计、服装材料、服装生产、服装营销和零售等领域的最新资料。同时，我们还为第 1 部分的文章配备了"Highlights（要点归纳）""For Review（复习巩固）""New Words and Expressions（生词与词组）""Translation（翻译练习）""Websites（相关网址）""Supplementary Reading（补充阅读）""Notes（难点注释）"和"Keys to Translation（翻译参考答案）"，以巩固读者对该部分内容的学习。其中需要读者重点掌握的翻译部分，在正文中是以斜体字出现的，并在"Translation（翻译练习）"中集中列出，读者可以在"Keys to Translation（翻译参考答案）"中查看参考答案。第 2 部分收录了服装生产和外贸方面的实用文章，内容涉及成本单、装箱单、设计师工作单、报盘与还盘、销售合同、信用证、提单和索赔、个人简历和招聘广告等。同时，我们也配备了"New Words and Expressions（生词与词组）"和"Notes（难点注释）"，以帮助读者学习该部分内容。

本教材选材新颖、内容丰富、实用性强，不仅能拓展读者的视野，提高学习服装英语的兴趣，还能使读者掌握标准的专业英语术语，学到更多有关服装设计、生产和营销的新理念，了解服装智能制造等相关新科技，提高专业水平。

本书的主编为北京服装学院服装艺术与工程学院的马凯副教授和语言文化学院的郭平建教授，编写者为王琪、张艾莉和张春佳副教授，并由吕逸华教授审校。在编著过程中，参考了大量教材及外文期刊论文，在此，对相关作者表示诚挚的感谢。

在本书的编写和出版过程中，得到了北京服装学院和中国纺织出版社有限公司有关领导、老师和编辑的大力支持和帮助，对此我们表示真挚的谢意。另外，由于我们水平有限，书中不妥之处敬请指正。

编　者

2021 年 2 月

教学内容及课时安排

章（课时）	课程性质（课时）	节	课程内容
第1章（6课时）	基础与训练（4课时）	A	时装业的演变
	应用与实践（2课时）	B	成本单
第2章（6课时）	基础与训练（4课时）	A	外观感知及其与服装的关系
	应用与实践（2课时）	B	装箱单
第3章（6课时）	基础与训练（4课时）	A	服装设计师做什么
	应用与实践（2课时）	B	服装设计师工作单
第4章（6课时）	基础与训练（4课时）	A	选择服装用织物
	应用与实践（2课时）	B	建立商业联系和索取样本
第5章（6课时）	基础与训练（4课时）	A	纸样制作
	应用与实践（2课时）	B	询盘、报盘和还盘
第6章（6课时）	基础与训练（4课时）	A	服装生产
	应用与实践（2课时）	B	销售合同
第7章（6课时）	基础与训练（4课时）	A	消费者需求与时装营销
	应用与实践（2课时）	B	信用证
第8章（6课时）	基础与训练（4课时）	A	时装零售商
	应用与实践（2课时）	B	提单和索赔
第9章（6课时）	基础与训练（4课时）	A	广告、专门促销活动和视觉销售
	应用与实践（2课时）	B	个人简历
第10章（6课时）	基础与训练（4课时）	A	时装产业的职业机会
	应用与实践（2课时）	B	招聘广告

注　该课程计划时数为64学时，其中讲授60学时，期中、期末测试各2学时。各院校也可根据自身的教学安排，对课时数进行调整。

目录
Contents

Chapter 1

Part A An Evolution of Fashion(时装业的演变) 2

        Highlights 6

        For Review 6

        New Words and Expressions 7

        Translation 7

        Websites 8

        Notes to Part A 8

        Supplementary Reading 9

Part B Costing Sheet(成本单) 12

Chapter 2

Part A Perception of Body Appearance and Its Relations to Clothing(外观感知及其与服装的关系) 18

        Highlights 24

        For Review 24

        New Words and Expressions 25

        Translation 25

        Websites 26

        Notes to Part A 26

        Supplementary Reading 28

Part B Packing List(装箱单) 31

Chapter 3

Part A What Does a Designer Do(服装设计师做什么) 36

	Highlights	40
	For Review	40
	New Words and Expressions	40
	Translation	41
	Websites	41
	Notes to Part A	41
	Supplementary Reading	42
Part B	The Designer Work Sheet(服装设计师工作单)	45

Chapter 4

Part A	Fabricating a Line(选择服装用织物)	48
	Highlights	52
	For Review	52
	New Words and Expressions	52
	Translation	53
	Websites	53
	Notes to Part A	54
	Supplementary Reading	55
Part B	Establishment of Business Relationship & Requesting Samples(建立商业联系和索取样本)	59

Chapter 5

Part A	Pattern Making(纸样制作)	62
	Highlights	66
	For Review	66
	New Words and Expressions	66
	Translation	66
	Websites	67
	Notes to Part A	67
	Supplementary Reading	68
Part B	Inquiry, Offer and Counter-Offer(询盘、报盘和还盘)	72

Chapter 6

Part A	The Apparel Manufacture(服装生产)	76
	Highlights	80
	For Review	80
	New Words and Expressions	80
	Translation	81
	Websites	82
	Notes to Part A	82
	Supplementary Reading	83
Part B	Sales Contract(销售合同)	87

Chapter 7

Part A	Consumer Demand and Fashion Marketing (消费者需求与时装营销)	94
	Highlights	98
	For Review	98
	New Words and Expressions	98
	Translation	98
	Websites	99
	Notes to Part A	99
	Supplementary Reading	101
Part B	Letter of Credit(信用证)	103

Chapter 8

Part A	The Fashion Retailer(时装零售商)	108
	Highlights	112
	For Review	113
	New Words and Expressions	113
	Translation	114
	Websites	114
	Notes to Part A	114
	Supplementary Reading	116
Part B	Bill of Lading & Claims(提单和索赔)	118

Chapter 9
Part A   Advertising, Special Events and Visual Merchandising(广告、专门促销活动和视觉销售)
           124

- Highlights   128
- For Review   128
- New Words and Expressions   128
- Translation   129
- Websites   129
- Notes to Part A   130
- Supplementary Reading   131

Part B   Résumé(个人简历)   133

Chapter 10
Part A   Careers in Fashion(时装产业的职业机会)   136

- Highlights   140
- For Review   141
- New Words and Expressions   141
- Translation   141
- Websites   142
- Notes to Part A   142
- Supplementary Reading   143

Part B   Recruitment Advertisement(招聘广告)   145

Keys to Translation   151
Glossary(词汇表)   160
参考文献   172

Chapter 1

An Evolution of Fashion

课题名称：An Evolution of Fashion（时装业的演变）

课题时间：6课时

训练目的：通过本章的学习，可以了解时装业的起源、发展和时装业发展趋势，培养学生对时装业发展的全局观；同时通过第二部分应用文的学习，了解服装成本单的内容和制作方法。

教学方式：教师主讲Part A的主课文和Part B的应用文，学生完成练习并课外阅读Part A后的Supplementary Reading（补充阅读）。

教学要求：1. 使学生了解时装业的历史。

2. 扩充学生服装方面的专业英文词汇。

3. 了解服装成本单的内容和制作方法。

Chapter 1
Part A

An Evolution of Fashion
时装业的演变

The world of fashion began with individual couturiers and evolved, as a result of the Industrial Revolution, into a mass-market industry. By studying this evolution, we become better equipped to understand the organization of the fashion industry and the directions in which it is moving.

The Beginning of Couture
高级时装业的开始

Although the history of fashion may be traced back hundreds of years, it was not until the late 1700s that individuality of design began to emerge. Styles were set by royalty and carried out by the dressmakers who served them. Only the upper class could afford what was fashionable and finely produced. The poor made their own clothing or wore the cast-offs of the rich.

By the end of the 18th century, one name had emerged in fashion design—Rose Bertin. Initially a milliner's apprentice, she became France's premier designer. As a result of the recognition she received from the Princess de Conti, Bertin was appointed court milliner in 1772. In that position, she was introduced to Marie Antoinette. She soon became the queen's confidante as well as her official designer. Eventually, Bertin became Minister of Fashion for the French court. As her reputation grew, she was commissioned to design hats and dresses for the aristocracy. Her fame spread to other countries, and she soon started to export her merchandise.

Garments for the wealthy class were elaborately tailored and trimmed. Each piece was hand sewn, embroidered, jeweled, and embellished to perfection. Aside from Bertin, the names of the dressmakers to the royal families and the aristocracy were generally unknown. Those who employed them jealously guarded their identities to avoid losing them to other families. [1]

During the early 19th century, the opulent designs that dominated the wardrobes of the rich began to disappear; less elaborate dress became the order of the day. It was not until after 1845, when the Englishman Charles Frederick Worth emigrated to Paris, that the world would come to know another designer. In Paris, he first worked for a fabric dealer, whom he convinced to open a dress department. In 1858, Worth was the first to open a couture house on Rue de la Paix. Along with a list of private clients in Europe and America, he was court dressmaker to Empress Eugenie of France. His success would soon motivate others to establish their own couture businesses.

Fed by the magnificent textiles and trimmings of nearby Lyon, it was natural for Paris to establish itself as the world's leading center for couture.

The Industrial Revolution
工业革命

Until about 1770, people worked in much the same manner as did their ancestors. Products were slowly made by hand. Cloth was hand woven, and a cobbler still used only a hammer, knife and awl to make a shoe over a last or form. [2]

During this time, the Western world witnessed the growth of the middle class, which prospered from new avenues of trade and industry, and spent money on such luxuries as fine clothing. As the middle class grew in importance, its members created new fashion directions. The business suit, for example, became an important element in a man's wardrobe. Before long, fine tailor shops were opened in London's Savile Row to provide this new business attire.

Changes, however, were taking place in the methods of production. In large part, they could be attributed to the growth of the textile industry, which was revolutionized by a series of timesaving inventions (Table 1.1). [3] In 1733, John Kay received a patent for his flying shuttle, which resulted in the manufacture of a loom that produced materials more rapidly. Similarly, spinning was a slow process until 1764, when James Hargreaves, a British spinner, invented the spinning jenny. He placed eight spindles on a frame, which could be turned by a single wheel. As a result, one spinner could simultaneously produce eight threads instead of producing one thread at a time.[4] Hargreaves later created a machine that could spin 16 threads at a time. Ultimately, even a child could run the machine and turn out work that had previously required 100 spinners. Then, in 1785, Edmund Cartwright invented the power loom, which wove cloth so rapidly that the handloom was quickly reserved for limited runs of special fabrics.

Table 1.1 Inventions of the Industrial Revolution that Changed Fashion

Year	Inventor	Invention
1733	John Kay	Flying shuttle
1764	James Hargreaves	Spinning jenny
1785	Edmund Cartwright	Power loom
1793	Eli Whitney	Cotton gin
1846	Elias Howe, Jr.	Sewing machine

The increased speed of the spinning machine resulted in demands for large supplies of cotton fiber. This problem was solved by an American, Eli Whitney. In 1793, he invented the cotton gin, which separated the cotton seed from the fiber so quickly and expertly that one man was able to turn out the work that once required 300 men.

Because of the competitive advantage these inventions gave to manufacturers, England was very protective of its discoveries, and forbade the emigration of textile workers and the exportation of its textile machines. Some workers, however, memorized the details of each machine's construction. These workers left England in disguise and were able to reproduce the machines in other countries. For example, Samuel Slater left England after learning the construction details for many textile machines. He opened a spinning mill in Rhode Island in 1790, where he introduced the factory system to the United States. During the Civil War, the demand for fabrics to manufacture uniforms helped the growth of US mills, most of which were in New England. By the end of the war, the mills were capable of mass-producing textiles. Fashion was now on the way to becoming a major industry in the United States, but one more step was necessary.

Although fabrics were being produced faster than ever before, it was not until the development of the first sewing machine that the world would be treated to a new generation of fashion.

Although Walter Hunt invented a sewing machine in 1832, he did not apply for a patent until 1854, when it was denied on the grounds of abandonment. On September 10, 1846, however, Elias Howe, Jr. did receive a patent for his sewing machine. As a result, he is generally regarded as its inventor. His failure to market the machine successfully led to attempts by others to further develop the machine. Finally, in 1858, Isaac Singer designed a machine that worked by the use of a foot treadle, thereby freeing the hands to manipulate the fabric. That year, the Singer Sewing Manufacturing Company was incorporated and sales reached 3000 units. *With this invention, women began to sew professional-looking clothes at home, and factories experienced the birth of ready-to-wear apparel.*

These new inventions created what is now as the garment industry, which includes textiles, manufacturers, retailers, licenses, franchises, fashion communications and market consultants.

The Future of Fashion
时装业的未来

At the turn of the 21st century, the world of fashion continues to change. Traditional rules of the game, which included faithfully following the dictates of specific designers on such issues as dress length, are finally being broken.[5] *Although globally renowned designers are still crowding the runways with outrageous styles at prices that only a few can afford, new designers are showing fashions that reflect what is taking place on the streets, in the political arenas, in the entertainment field, and in movements to protect the environment.*

Technology had risen to new heights. Every segment of the fashion industry, from the production of raw materials to final distribution to the consumer, takes advantage of ever-improving technological discoveries. The most notable are the CAD (computer-aided design) systems, which eliminate the need for endless paper patterns and time spent by the designer creating them at the drafting table. Other electronic applications, however, are moving the fashion industry into areas that not long ago seemed like fantasy.

For example, providing on-line computer services has enormous implications for fiber producers, end-product manufacturers, market consultants, retailers and promoters of fashion.[6] The Internet has gained prominence as a vehicle for

communicating, advertising, selling to suppliers and customers. Information about major retailers is available on-line. Individuals can download pictures of merchandise over their modem lines and place orders or ask questions via e-mail.

Another time saving and money saving invention is the fax machine, which disseminates information in a matter of seconds, enabling a design that originates in one country to be quickly copied in another. By transmitting the design electronically, the time and effort needed for traditional transmission is eliminated.

No one can predict the future of fashion— who the major players will be or how successful certain styles will become. By examining the past, one may begin to understand what this industry is all about and the number of variables that interact in the market place.

Many people— some famous, some unknown—impact the fashion industry. For example, anyone interested in fashion knows the names of such contemporary designers as Donna Karan, Ralph Lauren and Calvin Klein, and such past legends as Chanel, Balenciaga and Dior, but who can name a well-known patternmaker, sewer or trimmer? Although the former group steals the fashion headlines, the designers alone do not create fashion.

Highlights

- The evolution of fashion began with the individual dress maker, who designed clothing for the nobility.
- Fashion reached its present state only after the invention of timesaving machines that made the mass production of ready-to-wear apparel practical.
- The fashion industry is composed of textile producers, manufacturers, retailers, media and market consultants.

For Review

1. Who was the first person in fashion design with a specific clientele?
2. What were the five inventions of the Industrial Revolution that led to mass production of apparel?
3. What are the most recent developments in the fashion game?

New Words and Expressions

couturier n. 高级时装设计师,女式时装店
couture n. 高级时装业
mass-market adj. 大众市场
milliner n. 女帽商
premier adj. 首要的;最初的
confidante n. 知己,密友
aristocracy n. 贵族,权贵阶级
merchandise n./v. 商品,货物;推销,经营
garment n. 服装
trim v. 装饰,镶边
embroider v. 绣花
embellish v. 美化,装饰
opulent adj. 富裕的,豪华的
elaborate adj. 详尽而复杂的,精心制造的
cobbler n. 鞋匠
awl n. 锥子
last n. 鞋楦
avenue n. 方法,途经;林荫大道
attire n. (文学用语)服装
flying shuttle n. 飞梭
patent n. 专利
spinning jenny n. 珍妮纺纱机
spindle n. 锭子,纺锤
spinner n. 纺纱工,传动齿轮
power loom n. 动力织机

cotton gin n. 轧棉机,轧花机
forbid v. (forbade, forbidden) 禁止,不许
emigration n. 移民
textile n. 纺织品;纺织原料
disguise n. 伪装
abandonment n. 放弃;拒绝
treadle n. 脚踏板
segment n. 组成部分
manufacturer n. 生产商
retailer n. 零售商
license n. 许可证,执照
franchise n. 产品经销特许权
dictate n./v. 命令,支配,摆布
runway n. 跑道,时装表演走道
outrageous adj. 不寻常的,骇人听闻的
arena n. 活动场所
distribution n. 分配;销售
paper patterns n. 纸样
drafting table n. 绘图桌
fantasy n. 幻想
fiber n. 纤维
prominence n. 突出,显著
advertise v. 广告,宣传
disseminate v. 传播
fashion forecaster n. 流行预测员

Translation

Translate the following sentences italicized in the text into Chinese.

1. The world of fashion began with individual couturiers and evolved, as a result of the Industrial Revolution, into a mass-market industry.

2. During the early 19th century, the opulent designs that dominated the wardrobes of the rich began to disappear; less elaborate dress became the order of the day.

3. During this time, the Western world witnessed the growth of the middle class, which prospered from new avenues of trade and industry, and spent money on such luxuries as fine clothing.

4. Because of the competitive advantage these inventions gave to manufacturers, England was very protective of its discoveries, and forbade the emigration of textile workers and the exportation of its textile machines.

5. With this invention, women began to sew professional-looking clothes at home, and factories experienced the birth of ready-to-wear apparel.

6. Although globally renowned designers are still crowding the runways with outrageous styles at prices that only a few can afford, new designers are showing fashions that reflect what is taking place on the streets, in the political arenas, in the entertainment field, and in movements to protect the environment.

Websites

By accessing these Websites, you will be able to gain broader knowledge and up-to-date information on materials related to this chapter.

Union of Needle Trades, Industrial and Textile Employees:

http://www.unitehere.org

Notes to Part A

[1] Those who employed them jealously guarded their identities to avoid losing them to other families.

那些裁缝的雇主们对他们的身份严格保密,唯恐别家把他们挖走。

[2] Cloth was hand woven, and a cobbler still used only a hammer, knife and awl to make a shoe over a last or form.

(当时)布都是手工织的,鞋匠还在利用锤子、刀具和锥子在鞋楦上制鞋。

[3] Changes, however, were taking place in the methods of production. In large part, they could be attributed to the growth of the textile industry, which was revolutionized by a series of timesaving inventions.

然而生产方式发生了变化,主要原因是一系列省时的发明给纺织业带来了革命性的变化,推动了纺织业的发展。

[4] He placed eight spindles on a frame, which could be turned by a single wheel. As a result, one spinner could simultaneously produce eight threads instead

of producing one thread at a time.

他在一个机架上安装了 8 个锭子,由一个轮子带动回转。因此,一个纺纱工可以同时纺 8 根纱线,而不是以前一次只能纺一根纱线。

[5] Traditional rules of the game, which included faithfully following the dictates of specific designers on such issues as dress length, are finally being broken.

以前的服装设计,即使像服装的长度这样的细节都要由特定设计师指定,而现在这种传统的游戏规则最终被彻底打破了。

[6] For example, providing on-line computer services has enormous implications for fiber producers, end-product manufacturers, market consultants, retailers and promoters of fashion.

例如,提供网上的服务对包括纤维生产商、成品生产商、营销顾问、零售商和服装推销商在内的整个服装业有着巨大的影响。

Supplementary Reading
Case study 1

Tommy Hilfiger

One of the most important designers and manufacturers in the fashion world today is Tommy Hilfiger. Although his early success as a designer of men's wear was heralded (称赞) by consumers and retailers alike, his fashion offerings across a range of product classifications promise to make increased presence in the 21st century.

Wherever one turns in a department store, the Hilfiger brand and logo seem to be there. Whether in men's wear, women's sportswear, or children's wear, or in wearable accessories, footwear, fragrances and a host of products for the home, the brand has great consumer appeal.

The Tommy Hilfiger Corporation manufactures many of its own products, but relies on licensing agreements for other products, such as Pepe Jeans USA and Tommy Hilfiger Canada. The company's products are globally marketed throughout the United States, Canada, Mexico, Central and South America, most European countries, Japan, China and many Far Eastern nations. Its marketing expertise has established the Hilfiger label as a life-style brand all over the world.

The company's goal is to supply a youthful energy to its products and to bring a fresh perspective to classic apparel and other products. Although most of the attention thus far has focused on the company's apparel line, Hilfiger does not plan to stop there. The excellent performance of his Bed and Bath collection has given Hilfiger the impetus(推动力) to expand this line of bed linens, towels and bath accessories into much broader collection of the products for home that includes tabletop items, wall coverings and furniture.

Marketing has played an integral role in the company's brand success. In addition to its sponsorship(赞助者) of athletes, the company also relies on other celebrity sponsorships, as well as runway shows, publicity endeavors that result in editorial coverage, a vast investment in advertising, visual merchandising that places the products on selling floors in Hilfiger-designed fixtures, and numerous personal appearances by Tommy Hilfiger himself. Whenever a Hilfiger store visit is announced, the crowds turn out in enormous numbers.

The Tommy Hilfiger Corporate Foundation provides a means for the company to give back to diverse communities that have generously supported its products. It supports a number of causes that strive to improve the quality of life for young people. For example, the Foundation is a leading fundraiser for the French Air Fund, which enables underprivileged(下层社会的) urban youths to enjoy summer in the country.

The company's plans suggest that the Hilfiger label will continue to grow and expand into just about every appropriate consumer product classification in the 21st century.

Questions

1. What is the goal of the Tommy Hilfiger Corporation?
2. What has played an important role in the company's brand success?
3. What has the company done in marketing?

Case study 2

Nancy Park

Nancy Park, a young South Korean designer, received positive reviews from the fashion press for her collections designed under the Bravo label. In just 4 years, Park worked her way up from design assistant to head designer. In that position, she has been responsible for the company's offerings for the past 2 years. With a flair(资质,天赋) that is both whimsical(异想天开的) and fluid, she is steadily gaining popularity in America. As a result, Bravo's competitors have made offers for her services. In addition, a financial backer has offered to finance Park in her own fashion business. Although Bravo has offered to match the salaries offered by the competitors, it would not give Park a merchandise line with her own name. Unable to come to terms with her employer, Park has submitted her resignation and is now weighing numerous offers being presented to her.

Didier Ltd., a dress manufacturer, has offered her the head designer at a guaranteed salary plus 20 percent of sales. Her name would be on a label that reads "Didier Ltd., designed by Nancy Park".

Signal Fashions is a company that manufactures under its own label but also produces lines that bear the names of well-known designers via licensing agreements. Signal has offered a contract to Park that would create a separate division that would have only her name on every label. She would have complete design control, a salary triple(三倍于) her already excellent income, and the potential for part ownership in the company if the line was successful for 2 years.

Invest Associates, a financial backer with investments in five well-known couture operations, has offered capital that would establish Park in her first-class fashion business. The deal promises to finance 2 collections, make available contracting facilities enjoyed by its other interests, and give Park 50 percent of the profits. The label would feature Park's name, and she would have ultimate control over the company's design approaches.

With time drawing near for a decision, Nancy Park is still weighing the merits of each offer.

Questions

1. What are the advantages and disadvantages of each job offer?
2. Which offer do you think Miss Park should accept? Why?

Part B

Costing Sheet
成本单

成本单,顾名思义就是对某种商品的生产成本进行估算的表单。一般情况下,在购买所需样品的布料、设计的服装达到一定数量(如100件)之后,样品裁剪师将会为每件服装的成本做市场评估,并计算出单件产品所需材料的数量。然后,生产部经理确定每种款式的加工制作成本价。最后,由会计师计算出每种款式的全部费用,以便经营者了解盈亏。这些程序都是通过一系列的单据进行控制的。

Sample 1

如 Table 1.2 所示,这是一家公司的成本计算单。

Table 1.2 Cost Calculation Sheet

SEASON									
	DATE	INITIAL	GARMENT		SAMPLE		STYLE No.		
CONTROL									
SUBMITTED			DESCRIPTION						
RECEIVED									
ESTIMATED									
ADOPTED									
PRICE GOODS			MATERIAL COST						
Source and Description	Mill No.	Width	Unit Cost	YARDAGE			COST		
				Estimate	Standard	Actual	Estimate	Standard	Actual
TRIMMING				TOTAL PIECE GOODS COST					
Item	Source	Style No.	Unit	Unit Cost	QUANTITY			COST		
					Estimate	Standard	Actual	Estimate	Standard	Actual

Continued

	Overhead % Factor	Earned Labor			Earned Labor+Overhead %		
		Estimate	Standard	Actual	Estimate	Standard	Actual
	TOTAL						
	ADD WASTE FACTOR						
	TOTAL TRIMMING COST						
MANUFACTURING COST							
Cut-mark							
Operation							
Finishing							
TOTAL MANUFACTURING COST							
TOTAL COST							
GROSS PROFIT CALCULATION							
		Estimate			Standard		Actual
Standard Unit Selling Price Less% Discount							
Standard Net Selling Price							
Gross Profit $							
Gross Profit %							
SKETCH：				REMARKS：			

New Words and Expressions

sample 样衣
style No. 款号
initial 用自己姓名的首字母签字于(某处)
submit 提交,递交
estimate 估价
adopt 采纳
description 品名
piece goods 布匹,匹头
mill No. 厂号
unit cost 单价
yardage 码数

trimming 辅料
overhead 房租、电费等管理费用
earned labor 工费
cut-mark 裁剪
operation 加工
finishing 后整理
gross profit calculation 毛利计算
discount 折扣
sketch 服装效果图;草图
remark 备注

Sample 2

Table 1.3 是一家公司对某一长裤款式的成本单,其中包含了各种面辅料的价格、单位数量等信息。

Table 1.3 COSTING SHEET FOR TROUSERS

STYLE: SHELL FABRIC:
SIZE: As original sample
QUANTITY:

NO.	DESCRIPTION	UNIT	UNIT CONS.	UNIT PRICE	AMOUNT	AMOUNT (INC LOSE)
1	Body Lining	m				
2	Binding	m				
3	Pocket Lining	m				
4	Non-woven	m				
5	Waistband Interlining	m				
6	Hook	m				
7	3# YKK Zipper	m				
8	Buttons 24L	m				
9	Threads	pcs				
10	Care Label	pcs				
11	Import Label	pcs				
12	Security Label	pcs				
13	Fabric Label	pcs				
14	Plastic Bag	pcs				
15	Hangtag Sticker	pcs				
16	Plastic Bag Sticker	pcs				
17	Spare Button Bag	pcs				
18	Hanger	pcs				
19	Carton	pcs				
20	Hanging Bar	pcs				
	TOTAL MATERIAL					
	CM (CUTTING +MAKING)					
	CUSTOM BOOK FEE					
	FABRIC IMPORT FEE					
	GARMENTS EXPORT FEE					
	SUB-TOTAL					
	PROFIT					
	BANK					
	FOB PRICE					

New Words and Expressions

shell fabric   面料
binding   镶边,绲边
non-woven   无纺布
waistband interlining   腰带衬布
care label   护理标签
import label   进口标签

security label   防伪标签
hangtag sticker   吊牌贴纸
plastic bag sticker   塑料袋贴纸
spare button bag   备用纽扣包
FOB price   离岸价格

Chapter 2

Perception of Body Appearance and Its Relations to Clothing

课题名称：Perception of Body Appearance and Its Relations to Clothing
（外观感知及其与服装的关系）

课题时间：6课时

训练目的：通过本章的学习，学生可以了解应该如何审美和感知外观，如何通过着装来改善外观形象，把握外观与服装的关系；通过第二部分应用文的学习，了解装箱单的写法。

教学方式：教师主讲 Part A 的主课文和 Part B 的应用文，学生完成练习并课外阅读 Part A 后的 Supplementary Reading（补充阅读）。

教学要求：1. 使学生了解外观感知与服装的关系。

2. 扩充学生服装方面的英文词汇。

3. 了解装箱单的写法。

Chapter 2
Part A

Perception of Body Appearance and Its Relations to Clothing
外观感知及其与服装的关系

Our appearance is the most apparent individual characteristic. Although we are taught that we should not judge others by their appearance alone, relying on appearance to guide personal decisions and social interactions is not only natural, but also inescapable. The body and the way it is clothed and presented is a primary medium of expression, for it makes statements on the condition of society itself.[1]

Few people have a perfect body. *Most people would like to improve their appearance with appropriate clothing by camouflaging their less desirable attributes and highlighting the more attractive aspects of their bodies.* In order to design garments to present the best image of the wearer, it is necessary to understand the perceptions of beauty, body attractiveness and body image, as well as how the perception of body appearance can be modified through clothing.

Beauty
美的含义

What is beauty? Are there properties processed by an object which count towards beauty in all cases and which are sufficient or necessary for an object to be judged beautiful?[2] *One school of thought is that "beauty is in the eye of the beholder", that individual attraction is a result of personal experience, cultural background and specific circumstances.* Naomi Wolf, in her book, *The Beauty Myth*, argues that there is no such thing as a quality called beauty, which exists "objectively and universally". Some modern philosophers also believe that there are no principles of beauty, although there is a rational basis for genuine judgement of beauty. They argue that it is always possible to find an object which can be judged to exhibit principles identified as those of beauty but which does not evoke a pleasurable response, and, conversely, there may be objects which are experienced as beautiful but which do not exhibit the identified principles. [3] Nevertheless, the assumption that beauty is just an arbitrary personal preference may simply not be true. It cannot explain the fact that even two-month-old infants prefer to gaze at faces that adults find attractive.

If there are universal principles of beauty, what are they? *Ancient Greeks believed that the world is beautiful because there is a certain measure, proportion, order and harmony between its elements*. For centuries, the Golden Ratio or Golden Proportion, a ratio of 1 : 1.618 has been considered as the perfect ratio for beauty. It can be seen in nature and is used for art and architectural design. Linguists discovered that, although the same sound may mean entirely different things in different languages, there is a universal grammar underlying the combination of the sounds. Similarly, it has been suggested by many philosophers that beauty stems from the relationship between the elements comprising the whole. Evidenced from the rhyme of music and poetry, philosophers in the 20th century realized that such beauty is likeness tempered with difference or the fusion of sameness and novelty. Modern psychologists and biologists have echoed such a claim. [4] They found that men and animals, exposed for some time to a particular sensory stimulus, prefer new stimuli which are slightly different from the one with which they are familiar. "The likeness tempered with difference" is pleasing to the classification process, which is important for biological survival. [5]

Body Physical Attractiveness
人体吸引力

The classical Greek body proportions have been widely considered as ideal for centuries. The Greek ideal male and female figures are shown in Fig. 2.1 and Fig. 2.2 respectively.

The various body dimensions are measured in the unit of head length. For both the male and female, the height is approximately seven and half head lengths, with the fullest part at the hipline and wrist level dividing the total length exactly in half. [6] The neck is about one-third the length of the head, and the shoulder line

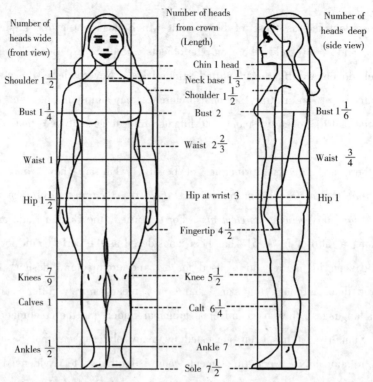

Fig. 2.1　Ideal Greek Proportions of Female Figure

Source: Horn M. J. and Gurel L. M. (1981). The Second Skin, Third Edition. Copyright (© 2003 by Fairchild Publications, Inc. Reprinted by Permission of Fairchild Books, a division of Fairchild Publications, Inc.)

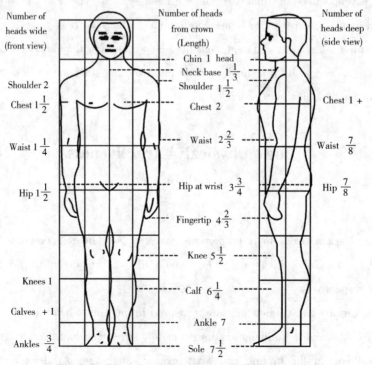

Fig. 2.2　Ideal Greek Proportions of Male Figure

Source: Horn M. J. and Gurel L. M. (1981). The Second Skin, Third Edition. Copyright (© 2003 by Fairchild Publications, Inc. Reprinted by Permission of Fairchild Books, a division of Fairchild Publications, Inc.)

slopes a distance of a half head length from the level of the chin. The fullest part of the bust or chest is located two head lengths from the crown. The waistline, which coincides with the bend of the elbow, is two and two-thirds of a head length from the crown. The knees are five and half head lengths from the crown and the ankles are seven head lengths from the crown. Male and female proportions differ only in circumference ratios. For the ideal female, the width of the hip frontal view is almost the same as the shoulder width. The shoulder width of the ideal male is greater than the width of his hips. *There is also a greater difference in the depth ratios from front to back in the female figure than there is in the male with respect to bust/ waist and waist/ hip relationships.*

Despite the wide appeal of the average Greek body proportions, the concept or perception of beauty ideals has never been static. It varies from time to time and from culture to culture. From the fifteenth to the seventeenth centuries in western cultures, a fat body was considered sexually appealing and fashionable. The ideal woman was portrayed as plump, big-breasted and maternal. By the 19th century, this had shifted to a more voluptuous, corseted figure, idealizing a more hourglass shape. In modern western culture, thinness coupled with somewhat inconsistent large breasts and a more toned, muscular physique has become the ideal of feminine beauty.[7]

In addition to historical factors, cultural differences play a significant role in the concept of beauty. For example, traditional Chinese culture associates plumpness with affluence and longevity, and Arab cultures associate greater body weight with female fertility.

Evolutionary psychology suggests that female physical attractiveness is based on cues of health and reproductive potential. Two putative cues to female physical attractiveness are shape (particularly the waist-hip ratio or WHR) and body mass index (BMI). Early researchers believed that a low WHR (i.e. a curvaceous body) corresponded to the optimal fat distribution for high fertility and hence female attractiveness. However, recent studies by Tovee and co-workers have shown that the body mass index (BMI), rather than WHR, is the primary determinant of female attractiveness.

Body Image
身体形象

The internal representation of one's own outer appearance, i.e. perception of one's own body, is termed body image.[8] Body image is important as it is strongly related to self-esteem and the development of personality attributes. A positive view of one's own looks may heighten one's self-esteem and lead to bold, successful interpersonal or business ventures, whereas a poor view of the physical self may weaken one's confidence. It has been found that a high percentage of women are dissatisfied with their body.

Modification of Body Appearance by Dressing
着装改善身体外观形象

Interaction between viewer, environment, body and clothing(观察者、环境、身体和服装之间的互动) *The appearance of the clothed body is a perception of the viewer (whether of the wearers themselves or others) in a social and climatic context.* It involves interactions between body, clothing, the viewer and environment (see Fig. 2.3). In mathematical terms, appearance is a complex function of body, clothing and environment (including social, cultural and other norms). Such a visual unit has been appreciated by DeLong, who defined the interactive unit as an Apparel-Body Construct. Viewing an Apparel-Body Construct is not just to scan and understand the visual components, such as line, shape, color, body shape, etc., which have their own meaning and expressive characteristics, but to perceive the contextual relationship between the components. Delong pointed out that the perception of clothing appearance is influenced by the Gestalt effect, that is, the whole is more than the sum of its parts. For example, the same jacket may appear different depending on what garments are combined with it.[9]

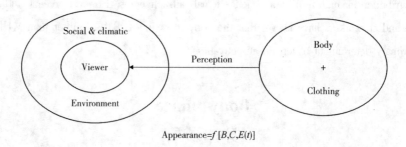

Fig. 2.3 Interaction Between Viewer, Clothing, Body and Environment

Changes in Body Cathexis
体形满意度的变化

Satisfaction with body appearance and its separate parts is termed as "body cathexis". Body cathexis is an evaluation of body and self-concept. A low value of body cathexis indicates dissatisfaction with one's own body appearance.

Body cathexis is highly related to satisfaction with the fit of the clothing. It has been reported that normal weight groups are most satisfied with their body and clothing fit. The overweight groups show much less satisfaction with their body and clothing fit. McVey found that ill-fitting branded garments, which are expensive and fashionable, give a message to consumers' that something is wrong with their body. However, less fashionable and less expensive private label merchandise does not carry the prestige to affect consumers' opinion of their own body. Markee et al. investigated the body cathexis of the nude body and the clothed body of 29 working women. They found that these working women were significantly more satisfied with their clothed bodies than with their nude bodies, showing the importance of dress in enhancing the perception of body appearance.

Illusion Created by Dress(着装引起的视错觉) The principles of illusion can be applied to the design of dress so as to camouflage the undesirable body attributes and to make the person's appearance closer to the ideal. Horn and Gurel have shown that, for a shorter figure with a sloping shoulder, the Muller-Lyer principle can be applied to create an appearance of increased shoulder width and body height (see Fig. 2.4). Design (a) in Fig. 2.4 can make the wearer look closer to the ideal proportion. A slender figure can be made fuller by adding fullness at the bust line and hipline and reducing the visual width of the waistline. A short figure can look taller by minimizing horizontal lines in the design. In general, parts of the body which are judged to be too large can be subdivided into smaller areas or counterbalanced by increasing the visual size of the surrounding elements. Body proportions that are considered too small may be masked or increased in size through the use of perspective and gradient techniques, or by minimizing the size of adjacent elements.

The perception of human faces may also be changed by hairstyles and collars.

For example, a round face may look better in a straight pointed collar and a square face may look better in a large collar to achieve the illusion of an oval face, which is the ideal in western culture.

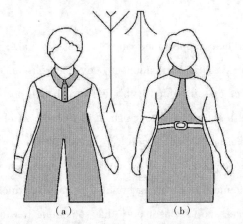

Fig. 2.4 Effect of Design on the Perception of Body Proportion

Source: Horn M. J. and Gurel L. M. (1981). The Second Skin, Third Edition. Copyright (© 2003 by Fairchild Publications, Inc. Reprinted by Permission of Fairchild Books, a division of Fairchild Publications, Inc.)

Davis summarized the visual design principles, such as repetition, parallelism, radiation, gradation, etc., and provided "recipe-style" guidelines for manipulating fabric texture, style, lines, decorative details, shape, form, color, pattern, etc., to achieve the desirable visual appearance.[10]

Highlights

- Philosophers believe that there are no universal principles of beauty, although there is a rational basis for genuine judgment of beauty.
- Despite the wide appeal of average Greek body proportions, the concept or perception of beauty ideals has never been static. It varies over time and from culture to culture.
- We can selectively modify our appearance through choice of clothing.

For Review

1. What is beauty? Do you think there are universal principles of beauty? Why or why not?
2. What factors affect the concept of beauty?

3. What is "body image"? Why do you think it is important?

4. Give some examples of the application of the principles of illusion in fashion design.

New Words and Expressions

perception   *n.* 感知,知觉
camouflage   *vt./n.* 伪装,遮掩,掩饰
attribute   *n.* 特征,属性
highlight   *vt.* 使……显著(突出)
modify   *vt.* 修改,修饰
myth   *n.* 神话
evoke   *vt.* 引起,唤起(记忆、感情等)
arbitrary   *adj.* 随意的,任意的,武断的
measure   *n.* (估价、判断事物的)尺寸;标准
temper(with)   *vi.* 调和或减轻某事物的作用,缓和
fusion   *n.* 熔化;合成,合并
sensory   *adj.* 感官的,感觉上的
stimulus(*pl.* stimuli)   *n.* 刺激
body dimensions   *n.* 身体尺度,尺寸

circumference ratios   *n.* 圆周长,这里指身体围度
voluptuous   *adj.* 性感的,体态丰满的
corset   *n.* 紧身胸衣
hourglass   *n.* 沙漏
physique   *n.* (尤指男性的)体格,体形
putative   *adj.* 公认的
cue   *n.* 暗示,提示
reproductive potential   生育潜力
index   *n.* 指数
function   *n.* 函数;功能
cathexis   *n.* 满意度
counterbalance   *vt.* 使平衡,抵消
gradient   *n.* 斜度,坡度
adjacent   *adj.* 相邻的,邻近的

Translation

Translate the following sentences italicized in the text into Chinese.

1. Although we are taught that we should not judge others by their appearance alone, relying on appearance to guide personal decisions and social interactions is not only natural, but also inescapable.

2. Most people would like to improve their appearance with appropriate clothing by camouflaging their less desirable attributes and highlighting the more attractive aspects of their bodies.

3. One school of thought is that "beauty is in the eye of the beholder", that individual attraction is a result of personal experience, cultural background and specific circumstances.

4. Ancient Greeks believed that the world is beautiful because there is a certain measure, proportion, order and harmony between its elements.

5. There is also a greater difference in the depth ratios from front to back in the female figure than there is in the male with respect to bust/ waist and waist/ hip relationships.

6. The appearance of the clothed body is a perception of the viewer (whether of the wearers themselves or others) in a social and climatic context.

Websites

By accessing this Website, you will be able to gain broader knowledge and up-to-date information on materials related to this chapter.

http://www.fashion.net

Notes to Part A

[1] The body and the way it is clothed and presented is a primary medium of expression, for it makes statements on the condition of society itself.

身体及其着装和外表是一种很重要的表达媒介，因为它反映出某一特定社会状况。

[2] Are there properties processed by an object which count towards beauty in all cases and which are sufficient or necessary for an object to be judged beautiful?

某种物体具有美的特征，是否这些特征在任何情况下都是美的呢？哪些特征是被认定为美的充分或必要条件呢？

[3] They argue that it is always possible to find an object which can be judged to exhibit principles identified as those of beauty but which does not evoke a pleasurable response, and, conversely, there may be objects which are experienced as beautiful but which do not exhibit the identified principles.

他们认为，往往有一些东西无论从哪一方面来说，都体现了所谓美的准则，但却并不能给人们美的感受。相反，有些能给人们美的体验的东西，却并不能体现出大家公认的美的标准。

[4] Evidenced from the rhyme of music and poetry, philosophers in the 20th century realized that such beauty is likeness tempered with difference or the fusion of sameness and novelty. Modern psychologists and biologists have echoed such a claim.

20世纪的哲学家从音乐和诗歌的节奏中得到启示，他们意识到，美即共性中存在差异，或者是新颖性和同一性的融合。现代心理学家和生物学家们都赞

Chapter 2 Perception of Body Appearance and Its Relations to Clothing

成这一说法。

[5] They found that men and animals, exposed for some time to a particular sensory stimulus, prefer new stimuli which are slightly different from the one with which they are familiar. "The likeness tempered with difference" is pleasing to the classification process, which is important for biological survival.

他们发现,无论是动物还是人类,当长时间受到某种感官刺激时,会倾向于选择与他们所熟悉的稍有不同的刺激。这种"共性中存在差异"有利于生物种类的进化,同时对于其生存也是非常重要的。

[6] The various body dimensions are measured in the unit of head length. For both the male and female, the height is approximately seven and half head lengths, with the fullest part at the hipline and wrist level dividing the total length exactly in half.

人体的各种尺寸都是以头部的长度为单位来测量的。无论是男性还是女性,人体的标准高度大约都是7个半头高,身高从臀围和腕线处正好一分为二。

[7] In modern western culture, thinness coupled with somewhat inconsistent large breasts and a more toned, muscular physique has become the ideal of feminine beauty.

现代西方文化中,理想的女性美主要体现在:身体瘦,乳房丰满,肌肉有力、发达。

[8] The internal representation of one's own outer appearance, i.e. perception of one's own body, is termed body image.

身体形象就是个体内心对自己外观的认识,即对自己身体的感知。

[9] Delong pointed out that the perception of clothing appearance is influenced by the Gestalt effect, that is, the whole is more than the sum of its parts. For example, the same jacket may appear different depending on what garments are combined with it.

狄龙指出,着装外观的知觉受格式塔效应的影响,即整体大于其组成部分相加的总和。例如,同样的一件夹克,由于与不同的服装搭配会显示出不同的效果。(Gestalt,格式塔理论,也叫完形理论,心理学术语,其主要观点就是:整体大于其组成部分的总和)

[10] Davis summarized the visual design principles, such as repetition, parallelism, radiation, gradation, etc., and provided "recipe-style" guidelines for manipulating fabric texture, style, lines, decorative details, shape, form, color,

pattern, etc., to achieve the desirable visual appearance.

为取得理想视觉效果,戴维斯总结出了重复、平行、辐射和渐变(即色彩呈渐次深浅原则)等视觉设计原则,并且提出了改进织物质地、风格、线条、装饰细部、形状、结构、色彩和图案等"万用款式"指导原则。

Supplementary Reading
Case study 1

Body Symbols and Dynamics

Much anthropological(人类学的) work has reported how various cultures use diverse body modification and adornment practices to mark individuals' sex and social status group memberships. Such practices involve both everyday symbolism (Vlahos 1979) and extraordinary situations such as in various rites(宗教仪式、典礼) of passage (Gennep 1908). Body decoration and maintenance in contemporary (当代的) American culture also reflect demographic criteria(统计指标) such as gender group, age, occupation and social class membership. On the other hand, individuals today have substantial freedom of self-expression; and simply knowing that someone is a certain age, earns so much money, and is a female will provide only a crude basis for predicting what she is inclined to wear, how she does her hair, or what kind of makeup she uses. Demographic data may suggest that certain broad personal appearance norms will or will not be operative, but demographics often provide little insight about why individuals select one product alternative over many available choices.

Marketers now largely accept the principle that to really "know" the consumer requires research beyond simple demographic profiles(统计图) of product users. This understanding has led to prolific psychological research streams that have investigated consumers' life-styles, motivations, fantasies and perceptual processing. The purpose of the study reported here is to investigate the relationship between individuals' purchase and use of various body modification and decoration products, and their own feelings about their bodies. The idea that there should be a discernible(分辨得出的) relationship between individuals' grooming(打扮,梳妆) behavior and their bodies relies on the fact that grooming is motivated parasomatic (辅助于肉体的)activity: it links psychic(精神的) with somatic(肉体的) realities. As such, grooming should reflect the dynamics of body self-esteem.

Chapter 2 Perception of Body Appearance and Its Relations to Clothing

The nature of these dynamics, and the direction of this relationship are points of some controversy. In their work *Body Image and Personality*, Fisher and Cleveland (1968) argue that the application of cosmetics serves as a compensation(弥补,补偿) mechanism for a deficient body self-image. Theberge and Kernaleguen (1979) explored this idea by testing the hypothesis that the greater the degree of satisfaction with the body, the less importance a person will ascribe to cosmetics. Body self-image was, in fact, found to relate positively and significantly to the total importance of cosmetics ($r=0.22$, $p=0.01$), and to the actual amount used ($r=0.18$, $p=0.01$). Theberge and Kernaleguen use these results to reject the compensatory function hypothesis. They suggest, rather, that cosmetics have a positive expressive function and that as satisfaction with the body increases, so does the importance of cosmetics. Other recent research has found evidence that females may be compensating for dissatisfaction with specific body features by buying more facial cosmetics and spending more time on daily cosmetic application (Cash and Cash 1982). Resolving these conflicting finding is difficult, because different dependent and independent measures are often used and sample characteristics may be dissimilar.

Questions

1. What problem do the demographic datum have according to theauthor?
2. What are the conflicting findings about the function of cosmetics?

Case study 2

Psychology and Comfort

Comfort, in the end, is the psychological(心灵的;心理的;精神上的) feeling or judgment of a wearer who wears the clothing under certain environmental conditions. Pontrelli developed a Comfort's Gestalt in which the variables influencing the comfort status of a wearer were listed comprehensively. The variables were classified into three groups: physical variables of the environment and clothing; psycho-physiological parameters of the wearer; and psychological filters of the brain. The Gestalt indicates that the comfort status of a wearer depends on all these variables and their interactions(相互影响,交互作用).

The psychology of comfort is the study of how the brain receives individual sensory sensations and evaluates and weighs the sensations to formulate subjective (主观的;主语的) perception of overall comfort and preferences which become our wear experience and influence our further purchase decisions.

Human perception of clothing and external environment involves all the relevant senses and has formed a series of concepts that we use to express these perceptions to each other. To understand the psychological processes we need to measure these perceptions in subjective ways. A subjective measure is the direct measure of the opinion of a person, which is the only factor of interest in carrying out the measurements. Since there are no physical instruments to measure what a wearer is thinking or feeling objectively, the only way to obtain the subjective perceptions is by use of psychological scaling. With psychological scaling, the process of making judgments is based on the scales of individual words or language that we collect from experience and share with peers throughout life.

Understanding how consumers perceive clothing and formulate their preferences is of compelling interest to both researchers and manufacturers. The overall sensory perception and preferences of a wearer to the clothing he or she wears are the result of a complex combination of sensory factors that come from the integration of inputs from various individual sensory modalities(形式;样式;感觉模式;modality 的复数) such as thermal, pressure and pain sensations. The individual sensory modalities are related to different mechanical-physical attributes of the garments. The sensory perceptions are also influenced by the psychological and physiological(生理学上的) state of the individual wearers and the external environment.

The process of integration is critical for developing an understanding of the psychological picture of clothing comfort. Subjective preference is further integration from inputs of the integrated sensory impressions in reference with past experiences, psychological desires, and the physiological status of the wearer, to form a final assessment of clothing. The integrated sensory impressions are highly related to the sensory factors that are derived from the latent pattern in various sensations. The relative contributions of the sensory factors to subjective preference may be different under different wear situations, since the psychological and physiological requirements of a wearer to clothing are dependent on specific combinations of the physical activities of the individual and the external environmental conditions.

Obviously, visual perception is probably the most important factor influencing aesthetic comfort of clothing. Kinesthetic(身体感觉;运动感觉) is the perception of body movement through the nerve endings that register the stretch or contraction of the muscle. Clothing can influence kinesthetic perception by providing restriction and pressure to muscle movement. For instance, tight-fitting(紧身的) Lycra cycling pants can enhance the kinesthetic perception in the muscles of the wearer's legs. Smell can be an important factor in comfort related to the body. Favorable smell from clothing may enhance the comfort perception of the wearer, while unfavorable smell may cause a feeling of discomfort. Occasionally, the sound generated from clothing influences the comfort sensory perception of the wearer. For instance, the sound of electric discharge during undressing a synthetic garment can enhance the discomfort perception caused by the electric insults on the skin. Probably, taste is the least important factor influencing clothing comfort.

Questions

1. What are the variables that influencing the comfort status of a wearer Comfort's Gestalt?
2. What perceptions are important for a body's comfort?

Part B

Packing List
装箱单

装箱单由包装车间填写。装箱单所填内容必须与信用证里标明的内容一致。一般包括:制造厂、装运唛头(标签)、发货号编号、合同编号、发货日期、装箱数、货物件数、重量及体积等。

Sample 1

下面是上海金海进出口有限公司的一张装箱单,货物要发往新加坡的AN-TAK发展有限公司。

PACKING LIST

TO: ANTAK DEVELOPMENT LTD.

INVOICE NO. SHGM70561
C NO. 00SHGM3178B
L/C No. 123456
Delivery Date Oct. 15, 2020

SHIPPING MARKS: ANTAK
00SHGM3178B
SINGAPORE
C/N: 1-190

C/NOS.	NOS& KINDS OF PKGS	QUANTITY	G. W. (KGS)	N. W. (KGS)	MEAS. (M^3)
	MEN'S COTTON WOVEN SHIRTS				
	1Pc in a poly bag				
	6Pcs in a kraft bag				
1~70	ART NO.: 1094L M L XL 3 3 4=10doz./cn	700 DOZ	2310KGS	2170KGS	9.8532 M^3
71~170	ART NO.: 286G M L XL 1.5 3 3.5=8doz./cn	800 DOZ	4500KGS	4300KGS	16.5816 M^3
171~190	ART NO.: 666 M L XL 1.5 3.5 3=8doz./cn	160 DOZ	660KGS	620KGS	2.8152 M^3
TOTAL		1660DOZ	7470KGS	7090KGS	29.25 M^3

SHANGHAI JINHAI IMP& EXP CORP. LTD.

New Words and Expressions

Invoice No. (Invoice Number) 发票号
C/NOS. (Carton Number) 纸箱号
L/C NO. (letter of credit) 信用证号
delivery date 发货日期
shipping mark 装运唛头（进出口货物包装上所做的标记）
PKGS (packages) 包装
poly bag 合纤袋

kraft bag 牛皮纸袋
Pc (piece) 件
ART (article) 货品，作品
DOZ (dozen) 一打（12件）
G. W. (gross weight) 毛重
N. W. (net weight) 净重
MEAS (measures) 体积

Chapter 2 Perception of Body Appearance and Its Relations to Clothing

Sample 2

Packing Specification

PART	CARTON NO.	CARTON SIZE	INDEX	COLOR	QTY OF BOX	QTY (PCS)/ CTN	QTY (PCS)	N. W. (kg)	G. W. (kg)
First part for Club Order	1	60 * 57 * 105	JT-001	Navy	1	10	10		
	2	60 * 57 * 105	JT-001	Navy	1	3	3		
	3	60 * 57 * 105	JT-001	Navy	1	0	0		
	4	60 * 57 * 105	JT-001	Navy	1	0	0		
	5~7	60 * 57 * 105	JT-001	Navy	3	0	0		
	8	60 * 57 * 105	JT-001	Navy	1	0	0		
	9~11	60 * 57 * 105	JT-001	Navy	3	0	0		
	12	60 * 57 * 105	JT-001	Navy	1	0	0		
	13~16	60 * 57 * 105	JT-001	Navy	4	0	0		
	17	60 * 57 * 105	JT-001	Navy	1	0	0		
	18~19	60 * 57 * 105	JT-001	Navy	2	0	0		
	20	60 * 57 * 105	JT-001	Navy	1	0	0		
	21~22	60 * 57 * 105	JT-001	Navy	2	0	0		
Second part for Club Order	23~25	60 * 57 * 105	JT-001	Navy	2	10	20		
	26	60 * 57 * 105	JT-001	Navy	1	1	1		
	27~30	60 * 57 * 105	JT-001	Navy	4	0	0		
	31	60 * 57 * 105	JT-001	Navy	1	0	0		
	32~39	60 * 57 * 105	JT-001	Navy	8	0	0		
	40	60 * 57 * 105	JT-001	Navy	1	0	0		
	41~49	60 * 57 * 105	JT-001	Navy	9	0	0		
	50~62	60 * 57 * 105	JT-001	Navy	12	0	0		
	63	60 * 57 * 105	JT-001	Navy	1	0	0		
	64~69	60 * 57 * 105	JT-001	Navy	4	0	0		
	70	60 * 57 * 105	JT-001	Navy	1	0	0		
	71~75	60 * 57 * 105	JT-001	Navy	5	0	0		
	76~78	47 * 46 * 85	TR-002	Grey	3	25	75		
	79	47 * 46 * 85	TR-002	Grey	1	2	2		
	80~86	47 * 46 * 85	TR-002	Grey	7	0	0		

Continued

PART	CARTON NO.	CARTON SIZE	INDEX	COLOR	QTY OF BOX	QTY (PCS)/CTN	QTY (PCS)	N. W. (kg)	G. W. (kg)
	Jacket for 1st part	60*57*105	JT-001	Navy			13		
	Jacket for 2nd part	60*57*105	JT-001	Navy			21		
	Trousers for 2nd part	47*46*85	TR-002	Grey			77		
	TOTAL PCS					111			
	TOTAL CARTONS					86			

New Words and Expressions

Pcs 件数

CTN (carton) 纸箱

navy 海军蓝的, 深蓝色的

Chapter 3

What Does a Designer Do

课题名称：What Does a Designer Do（服装设计师做什么）

课题时间：6课时

训练目的：通过本章的学习，学生可以了解服装设计师设计服装的全过程，包括研究色彩和织物、创造款式、开发产品、试销、打板和销售以及培养品位；通过第二部分应用文的学习，了解设计师工作单的写法。

教学方式：教师主讲 Part A 的主课文和 Part B 的应用文，学生完成练习并课外阅读 Part A 后的 Supplementary Reading（补充阅读）。

教学要求：1. 使学生了解服装设计的过程。

2. 扩充学生服装方面的英文词汇。

3. 了解设计师工作单的写法。

Chapter 3
Part A

What Does a Designer Do
服装设计师做什么

What exactly does a designer do? No two designers will answer this question in the same way because each designer does so many different things. As a rule, designers work for a wholesale apparel house or manufacturer, and their duties include a range of interrelated jobs.

Researching Colors and Fabrics
研究色彩和织物

The first step in creating a new line is to research fashion and consumer trends. Generally, the designer will begin by investigating color trends. *Information on color trends for the coming season usually comes from fiber companies and professional color services like Pat Tunskey.* On the basis of these general predictions, the designer selects the colors he or she feels will make the line unique and salable. Once the color story is set, the designer begins to review textile lines.

Creating Styles
创造款式

Once fabrics have been selected, the designer begins to create styles for the line. *The refabrications and versions of good bodies from the last season may provide the basic styles or staples of the new line.* Then the designer begins styling the new garments.

Aesthetic Appeal(美感) The garment should be attractive to the specific customer for whom it was created. Customers' aesthetic requirements differ with every size, price and age range, but everyone is looking for a stylish garment. Therefore, the fabric in every garment should be attractive and fashionable in print and color. [1] *Most customers touch a garment immediately after being visually attracted to it, so the hand (feel) of the fabric is as important as its appearance.*

Price(价格) The garment should obviously be good value for its price. If the price is too high, the customer will probably not try it on, even if she likes the style. *When the designer styles a garment, he or she must consider the price of every detail from the initial cost of the fabric to trims and construction methods.* Ultimately, the garment's success or failure may depend on cost.

Timing(及时性) A design should both fit into a general fashion trend and satisfy the customer's desire to be unique and fashionable. Again, how these requirements are met varies greatly for different customers. During the same period, the young junior customer may be buying authentic work overalls, the young missy (contemporary) customer may be looking for copies of European ready-to-wear, and the designer customer may be looking at Paris couture. [2] Many excellent products have hit the sale racks because they were too early or too late for the fashion taste of the moment. [3] The designer must be aware of general fashion trends and, more importantly, the trends that are influencing his or her particular market.

Fit(合体) Once the customer has selected a garment to try on, the chance of selling that garment improves. Now the way the garment looks on the customer's figure is crucial. The customer wants the garment to fit and make her look taller and slenderer. If the designer has created a garment to fit a slender, tall model, many average customers with figure problems will be disappointed when they try it on. The garment that conceals figure problems, flatters the face and body, fits well, and is pleasingly proportioned will be a success and success means a garment that sells! [4]

Care and Durability(保养和耐用) Most customers will examine a garment to see if it is well made. Easy care, particularly wash-and-wear, is an important part of making a satisfactory garment, and a satisfactory garment is one that performs well and looks beautiful. If the designer chooses a wash-and-wear fabric but lines the garment with a fabric that must be dry-cleaned, this defeats the fabric's purpose. [5] Many chain stores recognize the importance of durability. A chain store may give the manufacturer specific instructions about how the garment should stand up to washing and ironing. *High-priced garments usually require dry-cleaning because they are made of delicate fabrics and there is little customer resistance to the maintenance cost.* The successful performance of garments also depends on the designer's care in selecting trims and findings.

Developing a Line
开发系列产品

The techniques used to develop a line vary. A designer must be able to draw a working sketch or croquis of the garment planned. This working sketch may be shown to the manufacturer during a discussion about which garments are to be made up. The working sketch is given to the assistant as a guide in making the pattern and shows the placement of construction lines. *The silhouette or shape of the garment is important because it tells the patternmaker how much fullness should be in a skirt, sleeve or bodice.* A working sketch is placed as identification on each pattern. It is also used on the cost sheet.

Merchandise
试销

After all styles are finished, usually right before the season and before the styles are shown, the line is appraised and merchandised. This is also called weeding out the potentially unsuccessful styles[6]. Many of the numbers may be discarded. If they cannot be fitted into one of the price ranges feature by the house, even some of the most attractive numbers may be discarded.

How a line is merchandised depends on the policy of the house. The designer may show the line to the company head, who will review each piece. The company head decides whether the consumer will be willing to pay the price that will make the garment a profitable item.

Pattern Development
打板

The designer's involvement in the development of the production pattern varies greatly from company to company. At one extreme, the designer may never see the final product because all production procedures are separate from the design room. At the other extreme, the designer may be responsible for making the production pattern and supervising the grading and marker making. The latter system is prevalent in small firms.

Distribution
销售

The sales department consists of the head of sales (also called a merchandise person) and the sales people who work under his or her direction. Often included on the staff is a show room girl. The show room girl is in charge of the show room and the models. She maintains the line in good shape, and keeps up the book of sketches and swatches. She compiles and records all orders placed, handles special orders and trim replacements, and manages other details relating to retail stores and their orders.

The designer will often begin a new season by presenting the line to the sales staff and explaining the styling and theme at a sales meeting. Often the designer will join the sales staff when they are working the line with a buyer who makes many purchases or who represents a prestigious store. Sometimes the buyer will arrange an in-store promotion when a designer visits. A newspaper advertisement or fashion show (called a trunk show if the merchandise shown is for the next season) will be planned to bring in more customers than usual. This gives the designer an opportunity to see the garments on a variety of customers.

Developing a Taste Level
培养品位

Buyers return to their established resources for a specific type of styling, so a designer may be successful at one house and unsuccessful at another. For a designer to be successful, his or her natural taste must find expression. A person develops a taste level by constant exposure to fine merchandise in stores, in publications and on the streets. Good taste is the ability to recognize styles that will appeal to a specific segment of the market. The beginner should try to develop a feeling for one kind of apparel, a kind he particularly enjoys creating. However, before the designer begins to concentrate on one area, he should gain experience in several different areas. Fortunately the beginner has a wide choice of areas in apparel design.

Highlights

- The designer is responsible for a range of interrelated jobs, including researching colors and fabrics, creating styles, developing a line, merchandising, pattern development and distribution.
- Designers work closely with sales staff and buyers when they are working the line.
- Good taste is the ability to recognize styles that will appeal to a specific segment of the market.

For Review

1. What is the first step for a designer in creating a new line?
2. How do you develop a new line?
3. Is the designer necessary in the process of pattern making?

New Words and Expressions

line　*n./vt.* 型,款式,轮廓(线),(商品品种的)系列;给(衣服)装衬里
wholesale　*n.* 批发
inspiration　*n.* 灵感
style　*n./v.* 式样,风格
refabrication　*n.* 重新编造,重复使用,重新构造
version　*n.* (同一种物品稍有不同的)样式,复制品,版本
staple　*n.* 大路货,主要产品
aesthetic　*n.* 美学
stylish　*adj.* 时髦的,漂亮的
authentic　*adj.* 真迹,原作的,真实可靠的

overall　*n.* 外衣;工装服
silhouette　*n.* 轮廓
ready-to-wear　*n.* 成衣
defeat　*v.* 使失败,使受挫折
placement　*n.* 安置,布局
patternmaker　*n.* 样板师
fullness　*n.* 丰满度
bodice　*n.* 女装紧身上衣;上衣片,大身
appraise　*vt.* 评定,鉴定
weed out　*v.* 清除(不合格的人或物),淘汰
grading　*n.* 放缩板,推板
swatch　*n.* 样品
prestigious　*adj.* 有声望的,受尊敬的

Translation

Translate the following sentences italicized in the text into Chinese.

1. Information on color trends for the coming season usually comes from fiber companies and professional color services like Pat Tunskey.

2. The refabrications and versions of good bodies from the last season may provide the basic styles or staples of the new line.

3. Most customers touch a garment immediately after being visually attracted to it, so the hand (feel) of the fabric is as important as its appearance.

4. When the designer styles a garment, he or she must consider the price of every detail from the initial cost of the fabric to trims and construction methods.

5. High-priced garments usually require dry-cleaning because they are made of delicate fabrics and there is little customer resistance to the maintenance cost.

6. The silhouette or shape of the garment is important because it tells the patternmaker how much fullness should be in a skirt, sleeve or bodice.

Websites

By accessing these Websites, you will be able to gain broader knowledge and up-to-date information on materials related to this chapter.

Elle: *http://www.elle.com*

Vogue: *http://www.vogue.com*

Notes to Part A

[1] Therefore, the fabric in every garment should be attractive and fashionable in print and color.

因此,每件衣服布料的花型和色彩都应当漂亮和时尚。

[2] During the same period, the young junior customer may be buying authentic work overalls, the young missy (contemporary) customer may be looking for copies of European ready-to-wear, and the designer customer may be looking at Paris couture.

在同一时期,青少年顾客很可能会买正统的工作服,年轻的女士可能会买欧洲成衣的仿制品,而高价位时装的消费者则会关注巴黎时装。

[3] Many excellent products have hit the sale racks because they were too early or too late for the fashion taste of the moment.

许多优秀的产品之所以没有热卖，主要因为他们的品位超前或落后于当前的时尚。

[4] The garment that conceals figure problems, flatters the face and body, fits well, and is pleasingly proportioned will be a success and success means a garment that sells!

一件衣服如果能够掩盖身材缺陷，突出脸部和身体的优点，合身，并且具有悦目的比例，就是一种成功的设计。而所谓成功的设计就意味着服装会卖得好！

[5] If the designer chooses a wash-and-wear fabric but lines the garment with a fabric that must be dry-cleaned, this defeats the fabric's purpose.

如果设计师为一件衣服选择了免烫面料，却用一种必须干洗的面料做里衬，这样就破坏了使用该面料的初衷。

[6] After all styles are finished, usually right before the season and before the styles are shown, the line is appraised and merchandised. This is also called weeding out the potentially unsuccessful styles.

当所有款式都完成后，通常在该季节之前，在这些款式还没有公开之前，会对它们进行鉴定和试销。这种情况也被称作淘汰潜在的不合格式样。

Supplementary Reading

Case study 1

The Design Room

The designer's primary responsibility is the creation of seasonal collections, but he or she is also head of a department and has executive functions. The designer is expected to take charge of the personnel in the design (or sample) room. The designer must cooperate with the sales force, the production department, and the merchandisers, so creating a line is a team effort.

The designer usually has a staff in the design room to help develop ideas and execute sketches for the line. The designer must coordinate all staff activities and make sure the flow of work is constant. The number of people employed in the design room depends on the volume of work. The more functions a designer can perform in the design routine, the more valuable that designer will be to the firm.

A typical design room staff is described in the following list. Each job will require at least one person, depending on the amount of work to be completed.

Assistant Designer(助理设计师) This person is usually a good patternmaker. The assistant may work on the flat, drape in muslin or combine the two methods, but he or she must make an accurate pattern that translates the designer's working sketch into an actual garment.

Sample and Duplicate Cutter(样品及复制裁剪师) The cutter uses the designer's first pattern to cut sample garments and duplicates out of sample fabric.

Sample Maker(样品制作员) This person constructs the sample garment and may press and hand finish the sample. The sample maker works with the designer and assistant designer to fit the sample perfectly to the model. Usually, the sample maker has had a great deal of experience in factory sewing methods and can spot potential difficulties.

Sketcher(绘图员) A sketcher may be employed to make sketches for the designer and illustrations for the final promotion of the line.

Miscellaneous Duties(身兼多职员工的职责) These include shopping for findings and trimmings, supervising the making of duplicates, running the design room in the absence of the designer, filling out cost sheets, and supervising the receipt and shortage of sample fabrics.

Free-lance Models(自由职业模特) Fit models try on samples so the designer can fit and evaluate the garments. Models are also used to show the salespeople and buyers the samples during sales meetings and market weeks. The fitting models also work with the production patternmaker to perfect the fit of stock garments.

Beginners are rarely hired as designers. The responsibilities a designer must assume are so great that any manufacturer with an investment at stake will require an experienced person to head the design department. Occasionally a small or struggling house will hire a novice, but, as a rule, the beginner is hired as an assistant to the designer. In this capacity, he or she will perform a variety of services, depending on talent and training. The different jobs undertaken in the design room are an excellent starting place for a student who wishes a career in designing.

Questions

1. Why does the author say that creating a line is a team effort?
2. What positions does a typical design room staff include?
3. Why are beginners rarely hired as designers?

Case study 2

Custom Design

The smallest manufacturing firm is the single designer or seamstress who creates garments for private customers. This person handles all the aspects of the business we have outlined. This is an awesome task for one person, even though that person may produce a limited amount of merchandise. Many designers prefer to work on a small scale because they are not hampered by the creative limitations imposed by mass production. Custom designers enjoy enhancing the appearance of clients directly without the barriers of a salesperson or store.

The custom designer should have a good grasp of patternmaking to have flexibility in creating designs. A knowledge of fitting is essential. Usually the custom designer pads out a commercial dress form to represent a client's figure. A good sketch artist can save many long hours of sales time by sketching garments for clients before they are fitted in the first muslin(平纹细布,细平布).

Purchasing fabrics is usually a problem for the small designer because he or she often does not purchase enough of one fabric to buy wholesale(批发,趸售). Good fabric stores will often sell small amounts of fabric at a discount to an independent designer because the designer will purchase a great deal of fabric over the year. In addition, a custom designer can obtain a resale(转卖;转售) number from the state in which he or she works that permits the purchase of materials at a discount and eliminates sales tax on items for resale.

The custom designer often hires a skilled seamstress to assist in producing actual garments. A good accountant(会计;会计师) is a great help to the independent designer and will often assist in determining a fair markup figure to charge so that the business will be profitable.

Sales are best promoted by word of mouth. A satisfied customer will bring friends to the designer who creates attractive garments. Some custom designers will participate in fashion shows for women's group. This is an expensive but effective way to solicit[索求,请求……给予(援助、钱或信息);征求] clients. Often people who are hard to fit in commercial garments make valuable return customers. Advertisements in local newspapers and magazines will also attract clients.

Questions

1. What is custom design?
2. What positions are needed for the custom designer?

Part B

The Designer Work Sheet
服装设计师工作单

服装设计师要记录、存档所有的设计式样,填写设计师工作单。这个工作单有助于生产部门计算成本和采购服装面料、饰物。下面就是一个典型的设计师工作单。

Sample

A Designer Work Sheet

DESCRIPTION: PRINTED DRESS				STYLE # 1625
				PRICE: $56.48
SIZE RANGE:	FIBER CONTENT:			SEASON: SPRING
CARE	COLOR:			DESIGN#
MATERIAL	YARD	PRICE	AMOUNT	
BODY: PRINT	4.82	2.85	13.737	
				(Picture)
TRIM:				
TOTAL FABRIC COST			13.737	
TRIMMINGS	QUAN	PRICE	AMOUNT	
BUTTONS-LINE#: 20 (covered)	6	0.04	0.24	
ZIPPER:				
BELTLOOPS:			0.03	
HANGER LOOPS:			0.02	
THREAD:				
OUTSIDE: FUSE			0.50	
PACKING:			0.09	
LABEL:			0.11	
QUALITY CONTROL:				
SHIPPING:				

Continued

TOTAL TRIM COST:	0.99
LABOR:	
CUTTING:	2.00
SEWING: $96.00 PER DOZ	8.00
PRESSING:	0.78
BONUS/INSURANCE:	0.47
GRADING & MARKING:	0.53
TOTAL LABOR COST:	11.78
TOTAL DIRECT COST:	6.507

New Words and Expressions

printed dress　印花连衣裙
care　保养
yard　码数
20 (covered)　20个包扣
belt loop　饰带,带襻
fuse　热压

hanger loop　吊襻
bonus　红利
marking　划样;排料
total fabric cost　织物总成本
total labor cost　合计人工成本
total direct cost　合计直接成本

Chapter 4

Fabricating a Line

课题名称：Fabricating a Line（选择服装用织物）

课题时间：6课时

训练目的：通过本章的学习，学生可以了解服装设计师在选择面料时应该了解和考虑的因素，如面料本身的特性以及市场的需求等；通过第二部分应用文的学习，了解在服装外贸中建立商业联系和索取样本信函的写法。

教学方式：教师主讲 Part A 的主课文和 Part B 的应用文，学生完成练习并课外阅读 Part A 后的 Supplementary Reading（补充阅读）。

教学要求：1. 使学生了解应如何选择面料。

2. 扩充学生服装方面的英文词汇。

3. 了解在服装外贸中建立商业联系和索取样本信函的写法。

Chapter 4
Part A

Fabricating a Line
选择服装用织物

  This chapter discusses the business of selecting fabrics and how the textile producer and manufacturer must work together to get textiles into the marketplace in appropriate garments. Textile science (the identification of fibers and textile constructions, dyestuffs, printing techniques and so on) is discussed only as it relates to the actual ordering of fabrics. Textile science is a complicated subject that should be covered in a class separate from one on garment design. Many fine textbooks are available on this topic and may be used as a reference to answer technical questions.

  The designer's responsibility in selecting fabrics for the line can vary quite a bit, depending on the manufacturer. For example, some designers are assigned fabrics to work on after management has selected all the piece goods for the season. Other designers are required to select and purchase both sample and stock yardage. Generally, the designer's responsibility falls between these two extremes. *Some designers review lines with either a stylist or the owner of the house, but most manufacturers encourage the designer to make aesthetic decisions about fabric.* Frequently, the manufacturer has a bad relationship with some textile firms, and the designer is discouraged from ordering their samples. Similarly, manufacturers with poor credit ratings are unable to purchase piece goods unless they pay cash.

  Timing yardage purchases is an important aspect of apparel manufacturing. *Generally, piece-goods salespeople encourage manufacturers who are buying substantial amounts of yardage to commit (buy a specific amount or a particular fabric) early in the season to ensure delivery at the promised time.* Manufacturers of fashion items try to wait for early sales results before purchasing fabric. The pressure from both directions—the salespeople wanting commitment for fabric and the production supervisor or owner wanting salable garments—often falls on the designer. [1]

  A very small manufacturer who has no credit history must purchase fabrics COD (collect on delivery) from a textile manufacturer or jobber. A jobber is a middleman who purchases large amounts of fabric from a mill or from other manufacturers and sells them in smaller lots at a slightly higher price. When a company is large enough to establish credit with its fabric suppliers, terms for stock yardage purchases are 60-net. This means that manufacturers do not have to pay for fabric that they have purchased until 60 days after the shipping date. [2] This allows the manufacturer to work with the textile company's money for two months. A well-organized manufacturer can cut, sew, ship garments to a store, and often receive payment during this time period. A manufacturer can ask a textile manufacturer for dating, which is an additional time period in which to pay a bill, to allow the manufacturer time to collect before payment of the principal amount is due. Interest is usually charged on the amount outstanding. Fabric costs are approximately one-third the cost of a garment. Contractors and factory workers that cut and sew the garment must be paid immediately. The money "float" afforded by 60-net payment terms for fabric makes it easier for a manufacturer to finance the business.

The designer usually reviews all fabric lines that have any relevance to the product, even lines that are above or below the typical price range. This wide sampling is important because the designer must know what competitors and other firms in the market are choosing. In other words, when designers thoroughly research all fabrics offered during a season, they will have an overview of all textile trends and innovations. [3] Fabrics beyond the usual price range are important because the designer can use an inexpensive fabric as a lining or an expensive fabric as a trim in a limited area.

Occasionally, the designer will find a new fabric that promises to be a good seller, in which case he or she will sample it and test the most suitable sewing methods by making it into a stock garment. When a new product is satisfactory, management will commit itself immediately for some amount of yardage but specify (assorts) colors and prints at a later time. New fabrics are usually in short supply at first, so an early order ensures that a manufacturer will have the first enhancement to ship that fabric to stores in the product area.

A designer needs to know who else has sampled a fabric and will try to avoid fabrics that have been chosen by competitors or those used in lower-priced lines. This is especially true for novelty and print fabrics. Staple fabrics may be used in competitive lines where styling determines success. [4]

When a designer finds a fabric to sample, a sample cut of 3 to 5 yards is

ordered. The salesman provides a color card and information on delivery dates for duplicate and stock yardage. Stock yardage is the large amount of fabric needed to cut garments ordered by stores. Stock yardage is usually purchased by a purchasing agent specializing in fabrics. If the designer anticipates dyeing or printing a special color, it is important to negotiate minimum yardage requirements.

A designer should see as many lines as possible. The piece-goods salespeople are well acquainted with events in the marketplace, and they often have information that can aid a designer in the choice of a fabric or print. One well-known designer in the California swimwear market saw each salesperson who called on her at least once a season. She sampled what she thought would be appropriate fabrics for her products. Because she saw everyone and rarely turned anyone away, this designer was called first whenever a salesperson had a new or different fabric. This gave her a tremendous edge over other local designers, who had discouraged salespeople at one time or another. *The cardinal rule for a designer should be this: see all the textile representatives you possibly can during a season; judge their products' relevance to the designs you are planning; and be flexible; yet practical.*

Designers should organize information on textile and trim resources as they shop the market each season. A small sketchbook can be tucked into a purse or briefcase. Label the book with the season, color ranges, the names of the firm and salesperson, and other information necessary to order sample and stock yardage. Other methods include a loose-leaf binder with individual pages for each resource. Copies of sample orders can be filed after the note page. Keeping accurate records cuts down on time when looking for replacement fabrics or researching future lines.

Fabric Selection
面料的选择

The designer selects samples on the basis of price, aesthetics, fashion and the fabric's suitability for the line. Frequently, a beautiful fabric cannot be used because it is too similar to another fabric in the line. The designer should look for different weights and textures, crisp fabric and goods for draping, thin blouse weights and heavier bottom weights. The fabrics in the line should be balanced between novelty goods and base goods.[5] Novelty fabrics include prints, fancy woven patterns, textured and fancy knits, and textured wovens. Base goods are fabrics in solid colors and traditional patterns that can be used in many different styles.

Some manufacturers use one fabric repeatedly. For example, a manufacturer may only sell denim garments. Others repeat the same fabric seasonally, such as sportswear houses that use base goods like wool gabardine every fall and lightweight crepe every spring. A T-shirt manufacturer uses cotton interlock exclusively, and a bathing suit manufacturer styles a line primarily from Lycra spandex knits. Dress manufacturers may specialize in prints on rayon crepe or silk-like polyester. Designers are faced with changing the garment by dyeing or printing the fabric differently, adding trims, or altering the fabric with special treatments and finishes.

Using the same piece goods has several advantages. The fabric does not need to be tested for shrinkage and durability. Production techniques have been established, and garments can be costed with past history as a guide.[6] The customer expects to purchase fabric, and there is a built-in market for the product. A relationship between the textile producer and manufacturer has been established, and delivery of stock yardage can be projected for greater efficiency. Finally, volume purchases often result in a lower price.

The fabrics in a line should have a price range. The lower-priced fabrics can be made into more complicated styles. Higher-priced goods should be reserved for simple silhouettes.

When a sample fabric works well when it is made into a garment and is accepted as an item or group in the line, the sample garment must be duplicated.[7] Duplicates (dupes) are extra sample garments that are sent to road salespeople and showrooms in other markets. In many companies, the designer is responsible for ordering the duplicate yardage and supervising the construction of the dupes. Extra yardage is also needed to make production samples. Each time the designer orders the fabric, he or she should check the delivery date and the price of the stock yardage. The delivery date is determined by the turn-time of the fabric producer. Turn-time includes the time it takes to knit or weave the fabric, dye and finish it, and ship it to the manufacturer. Turn-time depends on many factors and can range from immediate delivery for yardage that is on hand in the producer's warehouse to several months. The designer should inquire about turn-time and relate it to the manufacturer's schedule.

Ordering of piece goods for stock cutting is usually handled by the manufacturer after the number of confirmed sales and the number of projected sales for the item have been calculated. A contract for the yardage purchase is drawn up between the fabric company and the manufacturer.

Designers based in regional design center frequently travel to New York to preview new fabric developments. They can discuss their ideas and color schemes directly with textile designers and technicians. Textile designers and principals also visit regional manufacturing areas to meet and work with designers.

Despite contracts and promises, production problems sometimes arise as textiles are produced in quantity. Unacceptable delays require that the fabric be replaced, and designers are often required to find the replacement. *Quick delivery of an acceptable replacement is critical, and designers often shop the fabric market looking for piece goods for the line in production as well as sampling new fabrics for the coming season.*

Highlights

- The designer must understand the processes of manufacturing apparel and producing textiles in order to coordinate the selection of fabric that can be delivered within the proper time frame to manufacture seasonal merchandise.
- The designer begins by sampling a fabric and testing it to see if it can be sewn and pressed in a particular design.
- A competent designer maintains an overview of the fabric market place to gain a competitive advantage and avoid missing an opportunity.
- The actual fabric selection depends on the price range, aesthetics and fabric mix appropriate to each line.

For Review

1. What is the designer's responsibility when fabricating a line?
2. Why should a designer see as many lines of fabric as possible?
3. What are the advantages of using the same piece goods for different products?

New Words and Expressions

assign   *v.* 分配；分派；指定　　　　　　credit rating   *n.* 信誉度
stock yardage   *n.* (产品的) 原料，库存备料　　commitment   *n.* 承付款项

jobber   n. 批发商;经纪人
terms   n. 期限;条款
due   adj. 应付的,到期的;约定的
outstanding   adj. 未完成的,未付款的
lining   n. 里衬
cardinal   adj. 重要的
binder   n. 活页夹,活页本
texture   n. 织物质地;(材料的)纹理,肌理
crisp   n./adj. 挺括(的),挺爽(的)
fancy   n./adj. 花式(的),时兴的纺织品（或服装）
denim   n. 粗斜纹棉布,劳动布;牛仔布

base goods   n. 基本商品,基本原料
gabardine   n. 华达呢
interlock   adj. 双罗纹的
cotton interlock   n. 棉毛布
Lycra   n. 莱卡
spandex   n. 氨纶
rayon   n. 粘胶长丝
crepe   n. 绉纱,绉布
polyester   n. 涤纶,聚酯
duplicate   n./v. 副本;备份
turn-time   n. 交货期限

Translation

Translate the following sentences italicized in the text into Chinese.

1. Some designers review lines with either a stylist or the owner of the house, but most manufacturers encourage the designer to make aesthetic decisions about fabric.

2. Generally, piece-goods salespeople encourage manufacturers who are buying substantial amounts of yardage to commit (buy a specific amount or a particular fabric) early in the season to ensure delivery at the promised time.

3. A designer needs to know who else has sampled a fabric and will try to avoid fabrics that have been chosen by competitors or those used in lower-priced lines.

4. A designer should see as many lines as possible. The piece-goods salespeople are well acquainted with events in the marketplace, and they often have information that can aid a designer in the choice of a fabric or print.

5. The cardinal rule for a designer should be this: see all the textile representatives you possibly can during a season; judge their products' relevance to the designs you are planning; and be flexible, yet practical.

6. Quick delivery of an acceptable replacement is critical, and the designers often shop the fabric market looking for piece goods for the line in production as well as sampling new fabrics for the coming season.

Websites

By accessing these Websites, you will be able to gain broader knowledge and up-to-date information on materials related to this chapter.

Fabric Online: *http://www.fabric.com*

Notes to Part A

[1] The pressure from both directions—the salespeople wanting commitment for fabric and the production supervisor or owner wanting salable garments—often falls on the designer.

布匹销售商希望尽快收到货款,生产总监或者老板希望自己的服装好卖,这两方面的压力通常都落在设计师的肩上。

[2] When a company is large enough to establish credit with its fabric suppliers, terms for stock yardage purchases are 60-net. This means that manufacturers do not have to pay for fabric that they have purchased until 60 days after the shipping date.

当一个公司大到足以跟布料供应商进行赊购买卖时,可签订60天后付购货款项的条款。也就是说,生产商可在货物起运60天之后付款。

[3] In other words, when designers thoroughly research all fabrics offered during a season, they will have an overview of all textile trends and innovations.

也就是说,当设计师对某一季所提供的所有布料进行了彻底的研究之后,就会对纺织品的流行趋势及创新有了一个全面的了解。

[4] Staple fabrics may be used in competitive lines where styling determines success.

具有竞争力的服装产品主要靠款式取胜,因此可以使用大路布料。

[5] The designer should look for different weights and textures, crisp fabric and goods for draping, thin blouse weights and heavier bottom weights. The fabrics in the line should be balanced between novelty goods and base goods.

设计师应当寻找不同重量和质地的面料。既要有挺括的,又要有垂悬性好的,一般女衬衣所需面料较轻薄,反之下身较厚重。而且应把握好新面料和常用面料之间的平衡。

[6] Production techniques have been established, and garments can be costed with past history as a guide.

一旦生产技术已经确定,服装就可以参考过去的生产情况进行成本核算。

[7] When a sample fabric works well when it is made into a garment and is accepted as an item or group in the line, the sample garment must be duplicated.

如果样料做成衣服后效果好,并被认可为可以规模生产的一种或一组产品时,样衣必须完全地加以复制。

Supplementary Reading
Case study 1

Fabrication

Fabrication is the process of selecting or creating a style for a fabric. The fabric that is suitable for a specific style depends on the characteristics of the fabric:

Surface interest—color, aesthetics, pattern, texture.
Weight—correct weight for wear requirements, season and construction details.
Fabric hand—correct stiffness for silhouette; fabric feels pleasant and drapes well.
Fiber—suited to the season, good performance and easy care; allergic reactions rare.

These four elements form the character of the fabric. They also dictate many of the limits on styling. The designer should be familiar with all types of fabric, just as a potter is familiar with the properties of many types of clay. The designer is a fabric sculptor(雕刻家), and the human body is the frame of reference.

The experienced designer can look at a piece of fabric or feel it and envision the type of garment it can be made into. This ability is developed by experimenting with many different fabrics and a great variety of styles. Usually, the designer has a working knowledge of basic textile construction, dyeing, and finishing and is familiar with natural and synthetic fibers. This basic knowledge can be expanded by talking with textile salespeople and reading trade papers. Innovations in the textile industry are constant, and fabric mills are eager to keep manufacturers and designers up-to-date on developments. Many textile and fiber firms attempt to solve construction problems that arise with a new fiber or method of construction.

The designer must evaluate the performance of a piece of fabric as well as the aesthetic(美学的) aspects before electing to use it in a line. A shrinking test under the pressing buck of the Hoffman press is the first step. To perform this test, cut a 12-by-12-inch piece of fabric and a corresponding piece of paper. Press the fabric, and then line it up with the paper pattern standard. If steam, heat and pressure have made the fabric shrink more than a half-inch in a 12-inch span, problems will occur when the garment is pressed. The average length of a jacket is 26 inches. A fabric that shrinks a half-inch on the pressing buck will shorten the jacket by 1 inch. Interfacing and lining usually do not shrink, so a jacket would be very difficult to construct in this fabric.

Some designers wash or dry-clean a fabric once it has been made into a sample to see how it performs. Trims should also be tested to make sure they can withstand the same care instructions as the base goods. The designer may elect to send questionable fabrics to a professional testing laboratory that will run tests for color fastness(染色牢度), washability and abrasion. These professional tests are especially important for garments that must stand up to the rigid quality-control inspections of mass merchandisers like Sears, JC Penney and Montgomery Ward.

Working with fabrics is an excellent background for a beginning designer. Design students often get jobs selling fabrics over the counter, and this exposure to a variety of fabrics is good training. Home sewing is another good way to learn about fabrics. Experimentation and observation are the keys to developing a sense of fabrication.

Questions

1. What are the four characteristics of a fabric?
2. What must designers do before they use a fabric in a line?
3. What are some good ways for a beginning designer to learn about fabrics?

Case study 2

Fiber Fashion Cycles

Fibers have fashion cycles just as silhouettes and colors do. Consumer preference changes, reflecting different lifestyles and values, though this is an evolutionary change that often spans a decade or more. Technical innovations also create changes in fiber demands. Polyester(涤纶) became popular in the early 1960s. High-fashion designers, such as Halston and Bill Blass, used the innovation ultrasuede(仿麂皮织物) and polyester jerseys to create expensive garments. At the same time, the younger generation discovered denim(粗斜纹棉布) jeans and jackets. Polyester continued in great demand for more than two decades, especially with the middle class, who valued performance and the easy-care characteristics of the synthetic fibers. Natural fibers were considered appropriate for the young, and denim was a counter-culture fabric.

Gradually, the market for natural fibers grew, and polyester declined in popularity. Many manufacturers did not identify these changes and went out of business. Finally, after decades of being relatively unpopular, polyester is returning to the fashion scene as microfilament and fleece and is fashionable once again. Evolutionary changes are often more difficult for business to adapt to because they happen over a long period of time and the urge is great to resist fundamental changes in a formula that has been successful for several years.

Fiber companies frequently offer advertising money to manufacturers who use substantial amounts of their fiber. The fiber company may contribute to an advertisement placed in a trade newspaper or magazine. Fiber money is also given to the manufacturer to be passed on to the retailer who advertises a specific garment to the public. The manufacturer may contribute money to the retailer because the manufacturer will gain if sales improve. These advertisements must carry the names of the fiber company and the manufacturer. Because advertising sells the product directly to the consumer, the advertising allowance is good for all three parties.

Questions
1. What factors do result in fiber fashion cycles? Give some examples.
2. In what ways are fiber companies, manufacturers and retailers related in advertising?

Case study 3

Smart Textile

Smart textiles are the textile version of smart materials. Traditional textiles are made from yarns that use materials chosen for their mechanical or structural qualities. Silk for its strength, light weight, and its affinity to dye; wool for its insulating ability; and polymers for their stretch, comfort, and price. The introduction of smart materials into textiles brings their inherent qualities to a flexible, wearable, and easily manufacturable product.

Smart materials have been around for years. The term "smart" or "intelligent" was first introduced in the US in the 1980s even though many smart materials had been in use for many years before then, but the introduction of smart materials to textiles is relatively new.

There are three categories of smart materials based on their functions: passive, active, and very smart materials. Each of these levels involves different types of technology. The lowest level of function is passive smart materials. They act as sensors, sensing the environment or stimuli. They gather information and can show what is happening on them such as color change, thermal or electrical resistivity. For example, a fabric that changes color when your body temperature changes. Photochromatic(光致变色的) inks are pre-programmed to trigger at a particular temperature and to change their hue. Exposure to UV light waves creates the reaction.

The next level of smart materials is active smart materials. These materials have the ability both to sense and to respond to external stimuli. When they are exposed to an environment, they act as both sensors and actuators(传动器). A number of active smart materials generate voltage when they are exposed to pressure, vibration, changes in pH, a magnetic field, or temperature. For example, when applying stress to a piezoelectric(压电;压电式;压电的) material voltage is created. A piezoelectric material is one that releases the same charge that is put into it. Because the reverse of this reaction is also true, when a voltage is applied to the material stress is created. This reaction has led to the development of materials that bend, expand, and contract when electrical current is applied.

Finally, very smart materials add a third function to the equation. These materials act as sensors and receive stimuli; they can react to information; and they can reshape themselves and adapt to environmental conditions. This category of materials is one of the most advanced and dynamic areas of research and discovery leading to new and exciting products and product categories: It includes shape-memory alloys(形状记忆合金), smart polymers, smart fluids, and other smart composites(复合物).

Questions

1. How to classify smart materials?
2. If you are going to design a style of protective apparel, do you want to use smart textile? Why?

Part B

Establishment of Business Relationship & Requesting Samples
建立商业联系和索取样本

要求与其他公司建立业务关系的信函主要包括:简述获得对方信息的来由;表达自己的意图与打算;介绍本公司的经营范围、基本情况、分支和联络处;介绍本公司的经营状况和商业信用并希望答复等。同时,针对具体的合作内容,买方要向卖方索取一定量的样品以确认是否满意。

Sample 1

A Letter for an Exporter Asking the Bank to Introduce New Customers

Dear sir,

We thank you for your cooperation for our business.

Now we hope to enlarge our trade in various kinds of apparel industrial products, but unfortunately we have no connections in California, USA.

Therefore, would you please introduce us to some of the most capable and reliable importers in the district who are interested in these lines of goods.

Your information will be greatly appreciated.

Yours faithfully,
Liang Wei
Managing Director
Export Department
China Lotus Apparel Ltd.

Sample 2

A Letter Requesting Samples

China Lotus Apparel Ltd.
88 Chaoyang Road, Beijing
The Peoples Republic of China
Tel: 86-10-85778294
Fax: 86-10-85778295
Jan. 9, 2021

Our Ref.: CS49/198101
Your Ref.: RMP/LW0714
American Apparel Trade Ltd.
P. O. Box No. 88, Seattle, Washington
The United States of America

Re: Woolen Dress Articles

Dear Sir,

　　We acknowledge receipt of your letter dated Jan. 9, 2021. We are interested in your woolen dress articles produced by the Desert Woolen Dress Group.

　　We shall greatly appreciate it if you will forward us some samples and relative pamphlets for our inspection.

　　Thank you for your attention to this matter. We are looking forward to your early reply.

Yours truly,
Liang Wei
Managing Director
China Lotus Apparel Ltd.

Chapter 5

Pattern Making

课题名称：Pattern Making（纸样制作）

课题时间：6课时

训练目的：通过本章的学习，学生可以了解长裤、短裤和裙裤的打板方法；同时通过第二部分应用文的学习，了解在服装外贸信函中询盘、报盘与还盘的写法。

教学方式：教师主讲 Part A 的主课文和 Part B 的应用文，学生完成练习并课外阅读 Part A 后的 Supplementary Reading（补充阅读）。

教学要求：1. 使学生了解应如何打板。

2. 扩充学生服装及外贸方面的英文词汇。

3. 了解在服装外贸中询盘、报盘和还盘的写法。

Chapter 5
Part A

Pattern Making
纸样制作

Trousers
裤子

Many trousers do not fit because of an inaccurate crotch length. To take this measurement, sit on a level chair and, with a tape measure, measure from waist, around hip curve, to chair. Add 2.5 cm (1 in) for ease. Make a note of measurement (Table 5.1). You will also need waist, hips, outside leg length and length from waist to knee. *Measurements in square brackets after each step are for the basic* 91.4 *cm* (36 *in*) *block.*

Table 5.1 Measurement of Trousers

Parts	cm	in
Waist: add 2.5 cm (1 in) ease	73.7	(29)
Hips: no ease	99.1	(39)
Crotch length	33.0	(13)
Outside leg	101.6	(40)
Waist to knee	58.4	(23)

Trouser Block 长裤(Fig. 5.1)

(1) *Approximately* 25.4 cm (10 in) *down from top of paper and* 20.3 cm (8 in) *in from left, draw a horizontal line AB, half the hip measurement* [49.5 cm (19.5 in)].

(2) *Square up to waist level* 20.3cm (8 in). *Draw across C to D* (Use individual measurement).

(3) From waist, measure down the crotch level and the outside leg measurement.

Complete the rectangle, noting waist, hip crotch and hemlines [33 cm (13 in), 101.5 cm (40 in)]. [1]

(4) Mark a central vertical line.

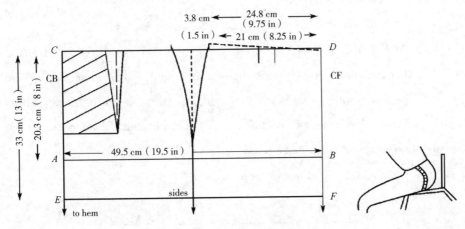

Fig. 5.1 Trouser Block

(5) At waist level, mark off 3.8 cm (1.5 in) each side of central line. Raise each point 1.3 cm (1/2 in). Draw in side-seam curves from raised point down approximately 15.2 cm (6 in). [2]

(6) On right, the front, mark a central point for pleat. Mark 2.5 cm (1 in) each side. Lower center front 1.3 cm (1/2 in) at waist, draw in curved waistline as for skirts. The pleat can be made into a dart and should be shorter than the back dart.

(7) Mark the central point of the back waist. *The dart size depends on the difference between amount in waist now and amount needed.* Measure the front, deducting amount in pleat [18.4 cm (7.25 in)]. Measure the back [21 cm (8.25 in)]. Add together [39.4 cm (15.5 in)]. Deduct half the required waist measure away from this [36.8 cm (14.5 in)]. The difference is approximately 2.5 cm (1 in). This is the necessary dart size.

(8) Mark a point 15.2 cm (6 in) below waist on central line. Mark 1.3 cm (1/2 in) points each side. Connect points to form the back waist dart. [3]

(9) Cut pattern on a horizontal line from center back to bottom of dart, and from waist down left dart line to same point without cutting through. To reduce the dart and create a sloping back seam, move left dart line over to central dart line (Fig. 5.2). Tape down. [4]

(10) Draw in the back curved waistline. Following (Fig. 5.2) extend E 10.2

cm (4 in), F 7.6 cm (3 in), A 2.5 cm (1 in), and B 1.3 cm (1/2 in). The crotch curves are drawn through these points. They are average measurements and should be adjusted to the individual. Corrections can be made to create the same body shape as the individual.

(11) At E draw a line bisecting the angle 5.7 cm (2.25 in).[5]

(12) At F draw in a 5.1 cm (2 in) bisecting line. Draw in curved crotch lines.

(13) For straight-leg trousers draw lines down from X and Y to hem line. These can turn out fairly wide, especially on larger hip sizes, but can be tapered to individual style. Most tapering is shaped from the thigh, but should be taken out equally between inside and outside seams on any style (Fig. 5.2).[6]

(14) Draw in knee level and short level.

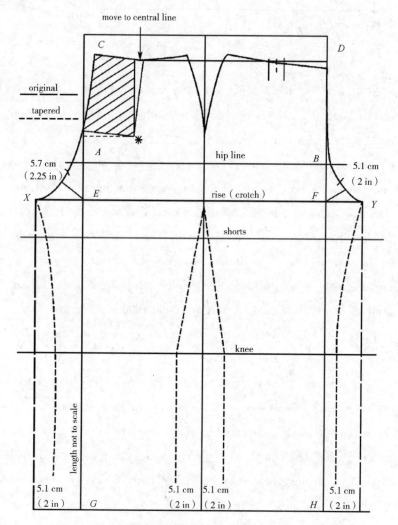

Fig. 5.2 Full Length Trousers

Shorts (短裤)

Shorts can be any length desired. They are often curved at the side for comfort. It is often a good idea to make up the trouser block as shorts as it could be less wasteful in the event of alterations. Always use woven fabric as knits give a false sense of fit (Fig. 5.3).

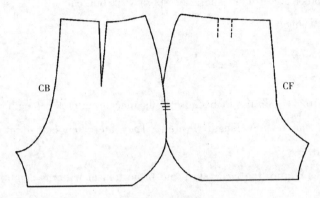

Fig. 5.3 Shorts

Culottes (裙裤)

Culottes require the basic skirt and basic trouser patterns combined. The solid line indicates the trouser pattern and the dotted line, the skirt pattern.

(1) Draw around the basic trouser pattern to the knee (front and back are designed in the same way). Position basic skirt as Fig. 5.4. Line up waist levels on CB and CF lines.[7]

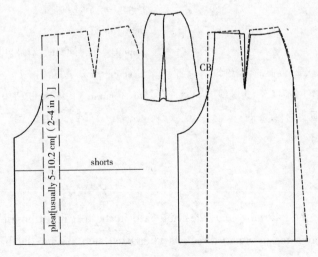

Fig. 5.4 Culottes

(2) *Trace the trouser pattern to the point where it touches the CB or CF skirt lines. Use skirt darting and waistline.*

(3) Add a pleat to the CB and CF skirt lines.

(4) *Re-position traced slack piece to the other side of pleat. Draw around complete culottes pattern.*

(5) Culottes are usually short for sportswear but can also be skirt length or Bermuda short length.

Highlights

- Many trousers do not fit because of an inaccurate crotch length.
- Shorts can be any length desired. They are often curved at the side for comfort.
- Culottes require the basic skirt and basic trouser patterns combined.

For Review

1. How is an accurate crotch length obtained?
2. Why is it a good idea to make the trouser block as shorts?
3. In making a culotte block, what kinds of patterns are required?

New Words and Expressions

crotch length　立裆
tape　*n.* 皮尺
ease　*n.* 松量
approximately　*adv.* 近似的,大约的
hemline　*n.* (裤边、袖口等)边缝线
vertical　*adj.* 垂直的,直立的
side seam　边缝

pleat　*n.* 褶,褶状物
dart　*n.* 省
sloping　*adj.* 倾斜的,有坡度的
bisecting　*adj.* 对角的
woven fabric　梭织物,机织织物
deduct　*vt.* 减去,扣除

Translation

Translate the following sentences italicized in the text into Chinese.

1. Measurements in square brackets after each step are for the basic 91.4 cm (36 in) block.

2. Approximately 25.4 cm (10 in) down from top of paper and 20.3 cm (8 in) in

from left, draw a horizontal line AB, half the hip measurement [49.5 cm(19.5 in)].

3. Square up to waist level 20.3 cm(8 in). Draw across C to D.

4. The dart size depends on the difference between amount in waist now and amount needed.

5. Trace the trouser pattern to the point where it touches the CB or CF skirt lines. Use skirt darting and waistline.

6. Re-position traced slack piece to the other side of pleat. Draw around complete culottes pattern.

Websites

By accessing these Websites, you will be able to gain broader knowledge and up-to-date information on materials related to this chapter.

Pattern Maket for Clothing & Apparel Industry Pattern Makers: http://www.apparelsearch.com/pattern_maker.htm

3D Fashion Design Software: http://www.clo3d.com

Notes to Part A

[1] From waist, measure down the crotch level and the outside leg measurement. Complete the rectangle, noting waist, hip crotch and hemlines [33 cm (13 in), 101.5 cm (40 in)].

向下量出从腰围到立裆水平线的长度和裤长。完成这个矩形,并标记出腰、臀、立裆和裤边缝线位置[33 cm (13 in), 101.5 cm (40 in)]。

[2] At waist level, mark off 3.8 cm (1.5 in) each side of central line. Raise each point 1.3 cm (1/2 in). Draw in side-seam curves from raised point down approximately 15.2 cm (6 in).

离腰水平中心线左右两边3.8 cm (1.5 in)各做标记点,然后分别抬高1.3 cm (1/2 in),从此点画出长约15.2 cm (6 in)的侧缝曲线。

[3] Mark a point 15.2 cm (6 in) below waist on central line. Mark 1.3 cm (1/2 in) points each side. Connect points to form the back waist dart.

从腰线处,沿中心线向下15.2 cm (6 in)做标记点,两边各分配1.3 cm (1/2 in),连接各点形成后片腰省。

[4] Cut pattern on a horizontal line from center back to bottom of dart, and from waist down left dart line to same point without cutting through. To reduce the

dart and create a sloping back seam, move left dart line over to central dart line. Tape down.

从后中线到省的底端剪开一条水平线,同时剪开左省线到省底端点,减少省量并做出一个斜后缝线,移动左省线到中心线,并用胶带纸固定好。

[5] At *E* draw a line bisecting the angle 5.7 cm (2.25 in).

在 *E* 点画出 5.7 cm (2.25 in)的对角线。

[6] For straight-leg trousers draw lines down from *X* and *Y* to hemline. These can turn out fairly wide, especially on larger hip sizes, but can be tapered to individual style. Most tapering is shaped from the thigh, but should be taken out equally between inside and outside seams on any style.

对于直腿型的裤子,从 *X* 和 *Y* 点做直线到缝边线。这会使裤子很宽,特别是对于大臀围尺寸,但是通过逐渐向下变窄也能做成适合个人的样式。多数可以从大腿处逐渐向下,但是任何款式内外侧缝都要等量裁剪。

[7] Draw around the basic trouser pattern to the knee (front and back are designed in the same way). Position basic skirt as Fig. 5.4. Line up waist levels on CB and CF lines.

画出裤子的基本板型到膝盖长度(前后片以相同方法设计)。按照图 5.4 所示放基本裙子的板型。在 CB 线和 CF 线上对齐腰线。

Supplementary Reading

Case study 1

Making and Appling Collars

Collars frame the neck and face and, as they are close to eye level, form one of the most noticeable parts of a garment. While there are a number of different styles, all collars fall into one of three main types: stand, flat and rolled. Plain, stand collars are the simplest ones to make and are frequently used on dresses and tops. Flat collars, such as Peter Pan and sailor's collars, are popular for blouses and also for children's clothes. Rolled collars and shawl collars are suitable for blouses, as well as for tailored jackets and coats.

Directory of Collars

Mandarin Collar(中式领) The pieces of this type of stand collar are often curved in shape at the front so that the collar slopes in smoothly towards the upper neck.

Shirt Collar(衬衫领) A shirt collar has an upright stand and a collar piece that folds down over the stand. The stand may be cut out as a separate band or cut in one piece with the collar.

Rolled Collar(翻领) On a rolled collar, part of the collar stands up at the neck edge. The rest folds back down. The rest folds back down. The stand section may be the same depth all around, or higher at the back.

Shawl Collar(围巾领) A variation on the basic rolled collar, the shawl collar has a stand section that gradually tapers down to a thin point at the center front.

Jabot Collar(胸饰领) In its basic form, this collar is a bias-cut square. The points of the square hang down in ruffles.

Stand Collar(立领) This type of collar stands up from the neckline seam. It can be made from a narrow band, or from a wider one that folds back on itself.

Flat Collar(平翻领) A flat collar sits almost flat at the neckline. Variations include Peter Pan and sailor's collars.

Parts of a Collar

Collars are made either from a single section of fabric and attached so that the ends meet at the center front or the center back, or from two sections of fabric so that the ends meet at the center front and the center back. Depending on the finished shape, a collar may be cut as one piece, which is folded in half, or as two pieces, which are seamed together.

Applying Interfacing

Interfacing is usually applied to the wrong side of the top collar piece of a rolled or flat collar, or to the outer part of a stand collar. It supports the top layer of fabric and also helps to mask the indentation of the seam allowances on the right side of the garment. The exception to this rule is on a tailored collar.

To A Very Lightweight Flat Collar Use sew-in interfacing on lightweight and fine silk fabrics, or on very fine fabrics use a matching organdie instead of interfacing. Pin the interfacing to the wrong side of the top collar piece, then tack in place.

To A One-piece Stand Collar Apply a lightweight sew-in interfacing to the whole collar piece. Take the interfacing lightly to the fabric along the fold-line (If using iron-on interfacing, interface the outer half of the collar up to the fold-line only).

To A Rolled Collar Apply sew-in interfacing to the wrong side of one-half of a one-piece rolled collar up to the fold-line. This half will form the top collar. Lightly herringbone stitch the interfacing to the fabric along the fold-line.

Questions

1. How many types of collars are there? What are they?
2. What are the functions of interfacing?

Case study 2

3D Printing

Beyond weaving and knitting, new manufacturing methods are helping create wearable materials that are not strictly textiles in the traditional sense. Designers are starting to use 3D printing in polymer-based(聚合物基的) materials to create wearable garments and accessories. 3D printing creates a three-dimensional object from a digital file through an additive process where the printer lays down successive layers of material until the object is created. 3D printers are capable of printing many different materials from plastics, paper, ceramic, glass and metals. Although this process is still in its infancy(婴儿期,幼儿期,初期), 3D printing will revolutionize the way we create and acquire products. The application of this manufacturing technique to wearable products is an exciting and evolving area of study.

ZAC Posen, GE Additive and Protolabs unveiled(展示,介绍,推出;将……公之于众) a collaboration featuring a range of innovative, sculptural 3D printed garments and accessories—inspired by the concept of freezing natural objects in motion —Met Gala(Metropolitan Museum of Art's Costume Institute in New York City, 纽约大都会艺术博物馆慈善舞会).

Over the past six months, Zac Posen and his creative team have explored a range of 3D printing and digital technologies with design engineers and 3D printing experts from GE Additive and Protolabs. With his vision and foresight(深谋远虑,先见之明), Zac Posen is demonstrating that almost anything is possible with 3D printing. "I dreamt the collection, GE Additive helped engineer it and Protolabs printed it," he said.

Four gowns and a headdress featuring 3D printed elements and structures were unveiled at the Met Gala, worn by British supermodel Jourdan Dunn, actresses Nina Dobrev, Katie Holmes, Julia Garner and Bollywood icon Deepika Padukone.

Deepika Padukone wore a custom Zac Posen metallic pink lurex(卢勒克斯金

属细线,卢勒克斯金属丝织物) jacquard (提花的) gown. This gown includes Zac Posen & GE Additive & Protolabs embroidery (绣花,刺绣技法,刺绣) which have been sewn on. The embroidery is made of Accura 5530 plastic and printed on a stereolithography (SLA,立体光刻,光固化) machine. These 408 delicately printed embroidery pieces are attached to the outside of the custom gown.

Jourdan Dunn wore a custom Zac Posen & GE Additive & Protolabs rose gown. The gown features 21 total petals, averaging 20 inches in size. The petals—each one of them unique— are fastened in place by a modular cage which is invisible from the outside. This dress was designed to a 3D recreation of Jourdan's body. The petals are finished with primer and color shifting (变换,更替,变动) automotive paint by DuPont (杜邦公司).

Nina Dobrev wore a custom Zac Posen & GE Additive & Protolabs bustier (抹胸). The bustier is a clear printed dress with 4-piece assembly for custom fit. The interior is designed to perfectly match Nina Dobrev's 3D re-creation. It is finished by wet hand sanding and sprayed with a clear coat to give it a glass appearance.

Katie Holmes wore a custom Zac Posen gown with a Zac Posen & GE Additive & Protolabs palm leaf collar accessory. The pearlescent purple palm leaves are draped over the shoulders and attached to the gown at the neckline. The structure is finished with pearlescent purple paint and holds the custom Zac Posen water coloured tulle gown at the clavicle.

Julia Garner wore a custom Zac Posen ombré silver to gold lamé gown with a Zac Posen & GE Additive & Protolabs headpiece. The intricate printed vine headpiece with leaf and berry embellishments is printed as a single piece and made of Nylon 12 plastic. The headpiece is finished by brass plating.

3D printing offers unique capabilities, such as near-complete design freedom, enabling the manufacture of designs that would have been difficult to achieve using other traditional methods of fashion design. GE Additive and Protolabs have worked closely together for a long time, including formally collaborating on development of 3D printing technology and production processes, so working together on this project was a natural fit, the partners say.

Questions

1. What is 3D printing, and what capabilities does it have?
2. Which style of 3D printing designed by Zac Posen do you like best in the Met Gala 2019, and why?

Part B

Inquiry, Offer and Counter-Offer
询盘、报盘和还盘

询盘就是买方想要获得某种产品的价格和其他交易的条款,向卖方发出信函或询价单(Inquiry)、征求报价单(Quotation)或邀请发盘(Offer)。询盘可以通过信函、电报、电传、传真和 E-mail 或者通过面对面的交谈进行。向老客户询盘比较简单,只要提及所需产品的名称和规格,向新客户询盘需要详细地提及产品名称、规格、质量、数量、价格条款(CIF、FOB 等)、付款条件(L/C、D/A 等)、交货时间、包装方式、折扣等,以便卖方能做出恰当的报盘。询盘信函力求文字简单、具体、合理、点到为止。可向卖方询问一般信息,索要产品目录单、价格表、样品、报价单等。

报盘是对询盘的答复,是卖方答复对方询问价格等情况的复函。报盘一般包括以下几个部分:向对方的询问表示感谢;列出货物名称、质量、数量及规格;包装条件;价格和优惠办法;付款条件;装船日期。

若买方对其所提出的条件持有异议并提出修改意见,其修改建议称为还盘。还盘应包括:感谢卖方的报盘;根据情况做出适当的还盘,即提出自己的建议;希望对方能接受并早日回复,或者提出其他促成交易的建议。

以下是对开襟羊毛衫(Cardigan Sweater)的首次询盘、报盘和还盘。

Sample 1

Inquiry

January 2, 2021

Dear Sirs,

  The Pacific Trading Co., Ltd. informed us that you are exporting a variety of cardigan sweaters. Would you please send us details of your various ranges, including sizes, colors and prices, and also samples of different qualities of material used?

  We are one of the oldest and largest department stores here and believe there is a promising market in our area for moderately priced goods of the kinds mentioned above.

  When replying, please state terms of payment and discounts you allow on purchase of quantities not less than five gross of individual items.

Yours sincerely,

(Signed)

Sample 2

Offer

January 4, 2021

Dear Sirs,

　　We are pleased to be told in your letter of January 2 that there are very brisk demands for our products, the cardigan sweater, in New York.

　　In compliance with your request, we are sending you our offer for your decision.

　　Commodity: Cardigan sweater in different color/pattern assortments.

　　Specification: Large(L), Medium(M), Small(S)

　　Packing: Sweaters are wrapped up in plastic bags and packed in standard export cardboard cartons.

　　Price: CIFC5% New York per piece in RMB

　　L: 788

　　M: 786

　　S: 781

　　Shipment: 3000 pieces each size per month, one month after receipt of L/C.

　　Payment: By confirmed, irrevocable L/C payable by draft at sight.

　　We trust the above will be acceptable to you and await with keen interest your trial order.

Yours truly,

(Signed)

Sample 3

Counter-offer

January 6, 2021

Dear Sirs,

　　Thank you for your letter of January 6, offering us cardigan sweaters of different assortments.

　　In reply, we regret to inform you that we find your prices much too high. Information indicates that other sellers in your neighboring countries are offering their products of the same kinds at much lower prices.

　　To step up the trade, we would suggest that you consider this prospective business in its long-term interests. It would be welcome if you could reduce your price by say, 15%. As the market is declining, we hope that suggestion is acceptable to you.

　　Looking forward to your early reply.

Yours faithfully,

(Signed)

New Words and Expressions

moderately priced 价格适中的
terms of payment 付款方式
gross 缩写为 gr., gro. 罗(计数单位,等于 144个,12打)
brisk 活泼的,活跃的
brisk demand 需求活跃
in compliance with 遵照
assortments 种类,类别
specification 规格
cardboard carton 纸箱
draft at sight 即期汇票
trial order 试订货
in reply 作为回答
step up the trade 促成交易
prospective 预期的,未来的
long-term interests 长远利益

CIFC 5%: CIF = cost, insurance and freight 成本、保险加运费
C = commission 佣金
CIFC 5% New York per piece in RMB 纽约抵岸价,包括 5%的佣金,以每件人民币(元)计算
confirmed, irrevocable L/C (letter of credit) 保兑的,不可撤销的信用证
category 类别,种类
prod line 产品线
target 目标
feedback 反馈
supplier 供应商
scarf 围巾
leather 羽绒
hosiery 针织品

Sample 4

Quotation List

Retail No.	Style Name	Pictures	Category	Prod Line	Retail Price	Color1	Total Qty	Target	Feedback	Comments	Supplier
001	Gloves		Gloves	Black label	299	Coffee as sample	2617				
				Black label	299		1813				
002	Scarf		Leather others Hosiery	Black label	199	Dark Brown as sample	3143				
				Black label	99		2114				
003	Belt		Gloves	Black label	299	Coffee as sample	2617				
				Black label	299		1813				
004	Sports Socks		Leather others	Black label	199	Dark Brown as sample					

Chapter 6

The Apparel Manufacture

课题名称：The Apparel Manufacture（服装生产）

课题时间：6课时

训练目的：通过本章的学习，学生可以了解服装公司的各部门以及各种服装生产的方式；通过第二部分应用文的学习，了解在服装外贸中销售合同的写法。

教学方式：教师主讲 Part A 的主课文和 Part B 的应用文，学生完成练习并课外阅读 Part A 后的 Supplementary Reading（补充阅读）。

教学要求：1. 使学生了解服装生产的各种方式。

2. 扩充学生服装方面的英文词汇。

3. 了解在服装外贸中销售合同的写法。

Chapter 6
Part A

The Apparel Manufacture
服装生产

The garment industry is characterized by both small and large manufacturing firms. *A businessperson with a good idea or the ability to sell a product can capitalize on his or her talent and pay small-business people to complete the manufacturing process.* In fact, almost all phases of production and selling can be contracted to outside firms. This saves the creative person from large investments in plant facilities and machinery. Also, a small apparel firm tends to have more styling flexibility and more merchandising innovativeness. Because the small manufacturer does not have to support a large factory and keep many machines busy, he or she can follow a trend quickly and then pull out as the fashion item saturates the market. [1] Often small firms specialize in servicing a small group of retailers who want exclusive, more expensive styling.

During the past dozens of years, many small apparel firms have developed into giants with sales in the millions of dollars. These large firms have departments that handle almost all the manufacturing, selling and promotional aspects of garment production. Frequently these firms have several divisions that produce noncompetitive lines. Some firms produce men's, women's and children's apparel.

Major Departments in an Apparel Firm
服装公司的主要部门

The average apparel firm is divided into three major departments: design, production and sales.

The Design Department(设计部门) It is headed by the designer, who is responsible for producing somewhere between 4 and 6 collections of garments per year: fall-winter, holiday, summer and transitional or early fall. Many progressive manufacturers avoid truly seasonal lines by working with a loose seasonal feeling and constantly adding and subtracting garments.

The Production Department(生产部门) It is responsible for mass-producing the line in various sizes and colors and filling orders placed by retailers.

The Sales Department(销售部门) *It markets the line produced by the design department and acts as an intermediary between buyers and the designer.* Small firms may combine the design and sales departments, especially if they specialize in copying current garments that are selling well.

In all three departments, the designer is actually a participant.

Contracting
承包生产

Contracting is hiring a factory or service to perform a specific part of the manufacturing process. The range of contracting services offered in an apparel-producing center spans all phases of production and design. The contractor is generally responsible for completing the work agreed upon by a specific time. Furthermore, the contractor must make the garments to a previously agreed upon standard.[2] Because the contractor assumes no responsibility for styles that do not sell, his or her risk is smaller than the manufacturer's.

There are both advantages and disadvantages to either contracting out work or maintaining an inside shop. To overcome the disadvantages, the manufacturer and the contractor must work closely together on the product and allow each other a fair share of the profits. The production person is the contact between the contractor and the manufacturer.[3]

Inside Shop(内部生产) Advantages: (a) greater quality control of products; (b) more accurate scheduling, special jobs more easily handled; (c) less physical movement of goods and personnel; (d) depreciation of facilities and machinery yields tax benefits.

Disadvantages: (a) large amounts of capital tied up in maintaining in terms of payroll, rent, repair and overhead costs, etc.; (b) during seasonal slow periods, workers have to be laid off, or an artificial workflow of promotional garments is created; (c) more time and effort devoted to union and employee demands.

Contractor or outside shop(承包商) Advantages: (a) great production flexibility when a contractor is hired only as production requires; (b) a highly paid technician, such as a patternmaker, does not have to be maintained during a period of no work; (c) no direct negotiations with labor or unions; (d) no capital investment or maintenance necessary; (e) easy entry into business.

Disadvantages: (a) less control over quality of product; (b) deliver and deadlines sometimes missed; (c) communication problems (many contractors speak foreign language); (d) extra physical movement of goods, resulting in greater chance of shortage or possible style piracy.

Foreign Production(海外生产) Manufacturers may produce merchandise in a foreign country. This is called offshore production. There are some advantages to this method because customs rates may be lower. *The manufacturer must carefully weigh the advantages of inexpensive labor against shipping costs, customs fees and the cost of coordinating the offshore production.*

Computer-Assisted Apparel Production
计算机辅助生产

Computers are revolutionizing the apparel industry. Medium-sized manufacturers can now afford to purchase systems to speed accounting and inventory control and to reduce production costs. Several factors have made this possible during the last dozen years. Technology has improved, and the price of sophisticated systems has been reduced gradually. Many companies now offer similar systems, and competition has helped to reduce prices. Foreign competition has forced domestic manufacturers to decrease production costs and reduce delivery time to capitalize on proximity to the market. *Increased flexibility resulting from the ability of these manufacturers to cut smaller lots and deliver rapidly allow stores to carry smaller inventories and makes an American-made garment more desirable.*

Importers also find computers assist in producing apparel offshore. They can design garments domestically and produce patterns on a computer system. When a garment sells well, it can be graded on the system and efficient markers made using accurate American technology. The domestic computer can transfer patterns overseas in a matter of minutes via a telephone hookup to a similar system. The garment can then be manufactured in a foreign country to take advantage of low labor costs abroad.[4]

Computer applications have been developed for almost every phase of the manufacturing process. Manufacturers traditionally begin to modernize their factories by purchasing a system that will grade and make markers and patterns. *When they find the savings are substantial, interlocking systems that will further speed production and reduce costs are added to the basic computer.*

Fabric and Design Applications(布料和设计的应用) Designers use graphic systems to design fabrics and garments. The design is drawn on an electronic tablet or photographed and transferred into the computer. Once the image is captured, the computer can change it at the designer's direction. A fabric print can be photographed, overlaid on a sketch of the garment and colored in many ways.

Pattern-making Systems(打板系统) The next step is to make a sample from the designer's sketch. The graphics system is connected to the pattern development system. The skilled pattern maker calls up the manufacturer's basic blocks (patterns) or a similar style and modifies them to produce the new style. The pattern is printed full scale on a specialized plotter and the sample is cut, sewn and fitted. The computer pattern is corrected, and a test marker is made to determine the amount of yardage needed for pricing purposes.

Grading and Marking Systems(推档和排料系统) After a garment has sold sufficient units to be mass-produced, a production pattern is made from the sample pattern stored on the computer. This garment will be corrected to fit the typical customer and graded from grade rules programmed into the computer in a few minutes.

Marker making is speedy and efficient on a computer, too. The computer can be trained to automatically analyze all the pattern pieces, compare them to previously made markers and create a new, efficient marker.[5] *System producers estimate that a minimum savings of 3 percent on fabric alone is typical for computerized markers, a significant factor since fabric is the most expensive element of a garment's cost.*

Automatic Cutting Systems(自动裁剪系统) Computers enter this phase of the manufacturing process by controlling the amount of fabric ordered. To do this, they compile the amount of orders written on each style. Arriving fabric is classified by color and shades within each color. Coordinated sportswear must be cut carefully so tops and bottoms going to one store are cut from the same dye lot. Computers assist in writing cutting tickets by matching orders to fabric on hand.

Garment Distribution on the Sewing Floor(缝纫车间的服装运送) Traditionally floor personnel carry bundles of partially sewn garments from one operator to another. Garment-moving systems carry the garments from machine to machine, speeding production time as garments are efficiently moved through the production process. Sewing engineers estimate time and motion savings of one-third.

Computerized Sewing Machines(计算机控制的缝纫机) Computerized sewing machines can be programmed to sew a specific number of stitches to perform a standard operation, like setting a zipper or sewing a man-tailored collar. The operator manipulates the fabric pieces in the machine, which automatically performs the programmed sewing operation.

Inventory Control and Shipping(存货清单控制和运送) Computerized inventory systems log in a garment as it is completed by the factory or contractor. Orders are tallied and then pulled for each customer according to the correct fabric color and shade. [6] Merchandise on hand is tracked quickly and efficiently, and the manufacturer always knows what is in the stockroom.

Highlights

- The major departments in an apparel-manufacturing firm are design (or product development), production and sales.
- Contracting is hiring a factory or service to perform a specific part of the manufacturing process, in order to minimize the risk to both contractor and primary manufacturer.
- All areas of garment construction now are being computerized, speeding design and production time and reducing the number of personnel required to make a garment.

For Review

1. What are the major departments in an apparel manufacture firm?
2. List the advantages and disadvantages of inside shops and outside shops.
3. How is computer technology applied in apparel manufacture?

New Words and Expressions

capitalize on   利用   facility   *n.* 设备,工具

flexibility *n.* 灵活性,机动性	sophisticated *adj.* 复杂的
merchandise *n.* 商品,货物	delivery *n.* 发货,交货
saturate *v.* 使饱和,浸透;使充满	proximity *n.* 接近,亲近
exclusive *adj.* 唯一的;高级的	hookup *n.* 连接装置,联网,连接
progressive *adj.* 上进的,进步的	substantial *adj.* 大量的;丰富的;充实的
subtract *v.* 减	interlocking *adj.* 连锁的,联合的
intermediary *n.* 调解者;中间者	graphic *adj.* 绘图的,图表的
specific *adj.* 具体的,明确的	tablet *n.* 写字板,书写板
span *v.* 横越,跨越	overlay *v./n.* 覆盖;覆盖图
assume *v.* 承担	block *n.* 原型,剪裁样板
depreciation *n.* 贬值;折旧	plotter *n.* 绘图机
artificial *adj.* 人造的,人工的	shade *n.* 色调
piracy *n.* 盗版,侵犯版权	stitch *n.* 一针,针脚,缝线
offshore *adj.* 海外的,国外的	log *v.* 记录[log in(计算机)输入指令开始]
accounting *n.* 会计学	
inventory *n.* 详细目录,存货	tally *n./v.* 记录;计算

Translation

Translate the following sentences italicized in the text into Chinese.

1. A businessperson with a good idea or the ability to sell a product can capitalize on his or her talent and pay small-business people to complete the manufacturing process.

2. It markets the line produced by the design department and acts as an intermediary between buyers and the designer.

3. The manufacturer must carefully weigh the advantages of inexpensive labor against shipping costs, customs fees and the cost of coordinating the offshore production.

4. Increased flexibility resulting from the ability of these manufacturers to cut smaller lots and deliver rapidly allow stores to carry smaller inventories and makes an American-made garment more desirable.

5. When they find the savings are substantial, interlocking systems that will further speed production and reduce costs are added to the basic computer.

6. System producers estimate that a minimum savings of 3 percent on fabric alone is typical for computerized markers, a significant factor since fabric is the most expensive element of a garment's cost.

Websites

By accessing these Websites, you will be able to gain broader knowledge and up-to-date information on materials related to this chapter.

Garment Contractors Association of Southern California:

http://www.garmentcontractors.org

SEAMS online:

http://www.seams.org

Notes to Part A

[1] Also, a small apparel firm tends to have more styling flexibility and more merchandising innovativeness. Because the small manufacturer does not have to support a large factory and keep many machines busy, he or she can follow a trend quickly and then pull out as the fashion item saturates the market.

而且,一个小型服装公司会有更大的设计灵活性和营销创新性。因为小型生产商不必要支撑一个大工厂和保持许多机器不断运转,他或她就能够紧跟时尚的潮流,当市场饱和时能及时脱身。

[2] The contractor is generally responsible for completing the work agreed upon by a specific time. Furthermore, the contractor must make the garments to a previously agreed upon standard.

一般来说,承包商负责在商量好的特定时间内完成工作。而且,制作的服装还要符合先前确定的标准。

[3] To overcome the disadvantages, the manufacturer and the contractor must work closely together on the product and allow each other a fair share of the profits. The production person is the contact between the contractor and the manufacturer.

为了克服其不足之处,服装生产商和承包商必须在产品生产上紧密合作,利益分配必须公平。生产者即是承包商和生产商之间的联系人。

[4] The domestic computer can transfer patterns overseas in a matter of minutes via a telephone hookup to a similar system. The garment can then be manufactured in a foreign country to take advantage of low labor costs abroad.

国内的计算机能够将板型在几分钟之内通过一个电话连接装置传到国外的一个相似的计算机系统上。这样就可以利用海外的廉价劳动力生产服装。

[5] The computer can be trained to automatically analyze all the pattern pieces,

compare them to previously made markers and create a new, efficient marker.

计算机可以通过编程自动地分析所有的样板单件,并把它们和以前制作的排料图进行对照,产生出新的、高效的排料图。

[6] Computerized inventory systems log in a garment as it is completed by the factory or contractor. Orders are tallied and then pulled for each customer according to the correct fabric color and shade.

计算机控制的存货系统可以记录工厂或承包商已经完成了的服装。同时它还能计算订单,然后根据正确的布料和色调与顾客沟通。

Supplementary Reading
Case study 1

The Case of the Production Dilemma

Jan Rogers and Peter English are considering forming a partnership(合伙关系) to manufacture moderately priced women's dresses. Each has had previous experience in the field. Jan was the production manager for Artway, a dress company that has been in business for 40 years; Peter was the assistant production manager for Bell Sportswear, makers of inexpensive skirts, pants and tops. Jan used in-house production and Peter out-side contractors.

For a designer, they have agreed on Renee Philips, who, while carrying the title of assistant designer, actually created many of her own styles. She is capable of preparing both the design and production patterns, and is familiar with computer-aided design technology.

The new company has limited financial resources, but the principals(负责人) want to see their dreams come true. They expect to begin their year with four collections, one for each season.

Together with Renee, the two are planning their production methodology. Jan believes the in-house approach is the most appropriate, whereas Peter favors the use of outside contractors. Renee is not certain which route would be most beneficial to the fledgling company.

Questions

1. What are the respective backgrounds of Jan Rogers and Peter English?
2. Which approach would you suggest the new company take: contracting or in-house production? Why?

Case study 2

Quick Response

Quick response, as outlined by the Management Systems Committee of the American Apparel Manufacturers Association, is "a management philosophy since it embraces actions by all functions of a business, working in concert with each other. It also involves working in concert with [和……相呼应(合作)] suppliers and customers in meaningful, in-depth trading partner alliances using uniform, standard procedures. The alliance has mutual objectives of increased sales and profitability and reduced inventory for all the partners." It is a combination of techniques that a business uses during all stages of production from the procurement(获得,取得) of raw materials to the delivery of the finished product to the consumer.

The goals of quick response include a reduction in production time, lowering of inventories, and increasing profitability. Two technologies that help achieve these goals are bar coding and electronic data interchange(EDI). They improve the communication process among manufacturers, materials suppliers and retailers. Bar coding has simplified the recording of point-of-sale(POS) information, which can be quickly sent to those businesses involved in shipping the goods from the production point to the consumption point. EDI has improved communication among all of these businesses. Because information can be conveyed quickly to everyone from producer to retailer, inventory replenishment(补给,补充) is fast and accurate. Ultimately, quick response results in smaller and more frequent orders, eliminating the need to overstock an abundance of goods. It provides the same advantage for all other segments of the fashion industry.

Question

1. What is "quick response"?
2. Why is "quick response" so important?

Case study 3

RFID: Tagging the new ERA

RFID(射频识别, Radio Frequency Identification 的缩写) is an automatic identification method consisting of several components such as tags, tag readers, edge servers, middleware, and application software. Among these the three important components are RFID tag (also known as transponder), RFID reader (also known as transceiver or interrogator) and software for dataprocessing. An RFID tag is a small object that can be attached to or embedded into a product, animal, or person. It consists of a tiny chip where the data is stored and an antenna to enable it to receive and respond to radio-frequency queries from an RFID transceiver.

In the field of Textile and Apparel the RFID is used in manufacturing, inventory control(存货控制,存货管理), warehousing(仓储), distribution(配送), logistics (物流), automatic object tracking, supply chain management etc. For example, the finished garments, different pattern pieces and accessories can be traced and the progress of the production process can be monitored. In processing and weaving, the fabric lots can be traced easily. In spinning mills, the bales(大包,大捆) of cotton and the yarns can be traced easily. The mixing of different yarn lots which is a major problem in spinning mills can be avoided.

RFID can be used in retail to monitor and control the floor level out-of-stocks (OOS, 缺货). It is mostly required where there is high product display density, low staffing level and chances of mishandling is very high. The stock level of the items is properly maintained and the items can be grouped according to their demand.

The theft of garments from the fitting room can be prevented by mounting a small reader at the entry point of the room. The tag information of the garment is captured when the customer takes the garment to the fitting room. The items that are taken to the fitting room but not come out are reported as potential loss items. So the RFID can be used to identify the missing merchandise and prosecution of the shoplifters.

Though the RFID cannot completely replace the barcode(条形码) technology, due to higher cost but the accuracy, speed and the return on investment is high in RFID system. The retailers, manufacturers and consumer goods companies like CVS, Tesco, Prada, Benetten, Wal-mart and Procter & Gamble now implementing the technology and exploring the impact of the technology on their business. The basic of success lies in understanding the technology and other features to minimize the potential problems.

Question

1. What is "RFID"?
2. What are the advantages of RFID?

Case study 4

Sustainability: Renewable Energy and Energy Conservation

Fabric batteries? Kicks that generate electricity? Solar energy-generating clothing? Scrunchable(可收缩的,可蜷缩的) antennae? All of these things sound like they are the product of an overactive imagination, but they are all in fact real products. Researchers are at work in laboratories around the world on projects that range from a device that will charge your cell phone as you run, to data that can be stored in your pocket, literally. Perhaps the most exciting aspect of this research, though, is its potential impact on sustainability, with proposals to generate energy from human kinetics(动力学) and environmental conditions such as sun, wind, and sound.

Researchers have now created a stretchable lithium-ion(锂离子) battery that is able to extend to four times its initial length with full recovery. This innovative product is completely flexible and continues working when it is folded, twisted, or stretched, making it the building block for solar power-generating clothes. Developed by an international team of engineers from China, Korea, and Illinois, the group see the SD (secure digital) card-sized battery being used in clothing (particularly activewear) that generates power, a flexible touch-sensitive(触敏的) skin for robots, temporary tattoos(文身) that monitor your vital signs(生命体征), and many other flexible futuristic wearable applications.

Even old-school(老派的,保守的) technology can be adapted to generate new ideas and results. At the University of Auckland in New Zealand, a group of researchers have adapted a simple, inexpensive rubber generator for a pair of shoes that can be used to harvest energy from walking. Over time, the movement of walking builds up to a watt of power—the amount used by your cell phone. The generator is made from artificial muscle material consisting of dielectric elastomer actuators (smart material systems that produce large strains), which compress and expand gradually, forming the charge. Each generator is estimated to cost under $4 to manufacture and would be an easy add-on for any number of clothing or shoe designs.

Water conservation is today a high priority, and globally we need to reduce water consumption. Around 780 million people lack access to clean drinking water; in the West an estimated 22 percent of household water is used for laundry alone.

Leading a group of researchers in China, Mingce Long of Shanghai Jiao Tong University, and Deyong Wu of the Hubei University for Nationalities, are working on developing clothes that clean themselves with sunlight and no water. The team drew on previous work that discovered a titanium oxide(氧化钛) solution able to remove stains in UV light. Since the use of UV light is not very practical, they turned their attention to sunlight instead. They developed a nano-particle(纳米颗粒) solution of titanium oxide in liquid form, into which they could then dip a cotton swatch. Once the swatch had been pressed and dried, the cotton was treated with a coating of silver iodide(碘化银) to enhance its light sensitivity. They stained the fabric with orange dye and exposed it to sunlight. The dye was completely broken down and, when tested, the cotton surface was free of bacteria. The team hope their discovery will lead to clothing that never needs to be washed, making laundry detergent(洗衣粉；洗衣液) a thing of the past.

Question

1. Why are the researchers working on the renewable energy?
2. How to improve a garment's sustainability? Try to give some examples.

Part B

Sales Contract
销售合同

销售合同是经过买卖双方协议后签订的具有法律效力的正式文件,用语必须准确、规范。合同包括正本(Original)和副本(Copy),通常采用的格式是三部分:约首(Head)、主体(Body)和约尾(End)。它们分别由以下内容组成。

约首:包括合同名称,合同编号,签约日期和地点,买卖双方的名称、地址和联系方式以及序言。

主体:主体是合同中最重要的内容,包含全部的具体条款:产品品名和规格(Commodity and Specifications)、产品品质(Quality)、产品数量(Quantity)、产品单价(Unit Price)和总值(Total)、装运时间(Time of Shipment)保险(Insurance)、包装(Packing)、运输标志/唛头(Shipping Mark)、保证条款(Guarantee of Quality)、检验索赔条款(Inspection and Claims)、付款方式(Terms of Payment)、运输方式(Terms of Shipment)、不可抗力条款(Force Majeure)、延期交货和惩罚

条款(Late Delivery and Penalty)和仲裁条款(Arbitration)。

约尾:包括有效期、份数、双方签名盖印以及备注等。

Sample 1

以下是一个英文销售合同的实例。

CONTRACT

Date: November 22, 2020   Contract No. : 2020-123

The Buyers: Rotterdam Textile Import and Export Company, Netherlands
The Sellers: China National Textiles Import & Export Corporation, Shanghai Fabric Branch

This contract is made by and between the Buyers and the Sellers; whereby the Buyers agree to buy and the Sellers agree to sell the under-mentioned goods subject to the terms and conditions as stipulated hereinafter:[1]

Name of Commodity: silk pajamas
Quantity: 1500 dozen
Unit price: $ 200 per dozen CIF 5% Rotterdam
Total Value: $ 300000
Packing: Each piece in a polybag, one dozen to a box and 15 dozen to a wooden case
Shipping Mark: Made in China
Country of Origin: China
Terms of Payment: By irrevocable L/C available by draft at sight[2]
Insurance: To be covered by the Sellers for 110% of the invoice value against All Risks and War Risk[3]
Time of Shipment: During January 2020
Port of Lading: Shanghai, China
Port of Destination: Rotterdam, Netherlands

Claims: Within 45 days after the arrival of the goods at the destination, should the quality, specifications or quantity be found not in conformity with the stipulations of the contract except those claims for which the insurance company or the owners of the vessel are liable, the Buyers shall, have the right on the strength of the inspection certificate issued by the C.C.I.C and the relative documents to claim for compensation to the Sellers. [4]

Force Majeure: The sellers shall not be held responsible for the delay in shipment or non-delivery of the goods due to Force Majeure, which might occur during the process of manufacturing or in the course of loading or transit. The sellers shall advise the Buyers immediately of the occurrence mentioned above the within fourteen days thereafter. The Sellers shall send by airmail to the Buyers for their acceptance a certificate of the accident. Under such circumstances the Sellers, however, are still under the obligation to take all necessary measures to hasten the delivery of the goods. [5]

Continued

> **Arbitration**: All disputes in connection with the execution of this Contract shall be settled friendly through negotiation. In case no settlement can be reached, the case then may be submitted for arbitration to the Arbitration Commission of the China Council for the Promotion of International Trade in accordance with the Provisional Rules of Procedure promulgated by the said Arbitration Commission. The Arbitration committee shall be final and binding upon both parties and the Arbitration fee shall be borne by the losing parties. [6]
>
> The Seller: The Buyer: (Signed)

Notes

[1] This contract is made by and between the Buyers and the Sellers; whereby the Buyers agree to buy and the Sellers agree to sell the under-mentioned goods subject to the terms and conditions as stipulated hereinafter:

兹经买卖双方同意按照以下条款由买方购进,卖方售出以下商品:

[2] Terms of Payment: By irrevocable L/C available by draft at sight

付款方式:不可撤销的信用证

[3] Insurance: To be covered by the Sellers for 110% of the invoice value against All Risks and War Risk

保险:由卖方按发票金额的110%投一切险和战争险

[4] Claims: Within 45 days after the arrival of the goods at the destination, should the quality, specifications or quantity be found not in conformity with the stipulations of the contract except those claims for which the insurance company or the owners of the vessel are liable, the Buyers shall, have the right on the strength of the inspection certificate issued by the C. C. I. C and the relative documents to claim for compensation to the Sellers.

索赔:在货到目的口岸45天内如发现货物品质、规格和数量与合同不符,除属保险公司或船方责任外,买方有权凭中国商检出具的检验证书或有关文件向卖方索赔换货或赔款。

[5] Force Majeure: The sellers shall not be held responsible for the delay in shipment or non-delivery of the goods due to Force Majeure, which might occur during the process of manufacturing or in the course of loading or transit. The sellers shall advise the Buyers immediately of the occurrence mentioned above the within

fourteen days thereafter. The Sellers shall send by airmail to the Buyers for their acceptance a certificate of the accident. Under such circumstances the Sellers, however, are still under the obligation to take all necessary measures to hasten the delivery of the goods.

不可抗力：由于人力不可抗力的缘由发生在制造、装载或运输的过程中导致卖方延期交货或不能交货者，卖方可免除责任，在不可抗力发生后，卖方须立即电告买方并在14天内以空邮方式向买方提供事故发生的证明文件，在上述情况下，卖方仍需负责采取措施尽快发货。

[6] Arbitration: All disputes in connection with the execution of this Contract shall be settled friendly through negotiation. In case no settlement can be reached, the case then may be submitted for arbitration to the Arbitration Commission of the China Council for the Promotion of International Trade in accordance with the Provisional Rules of Procedure promulgated by the said Arbitration Commission. The Arbitration committee shall be final and binding upon both parties and the Arbitration fee shall be borne by the losing parties.

仲裁：凡有关执行合同所发生的一切争议应通过友好协商解决，如协商不能解决，则将分歧提交中国国际贸易促进委员会按有关仲裁程序进行仲裁，仲裁将是终局的，双方均受其约束，仲裁费用由败诉方承担。

Sample 2

以下是另一个英文销售合同的实例。

AGREEMENT

DT：20-05-2021

Between

BIFT Fashion Group (Beijing) Co, Ltd
2 Yinghua Road, Chaoyang District, Beijing
100029, P. R. China
&
ANT DENIM TECHNOLOGY LTD.
NEWYORK

This agreement covers on style(s) as followed：
Men's Woven jeans
COFFS JEANS (style name) unit price USD 13.00/9907 Pcs/total USD128791.00

Continued

The reference number is 200 with planned shipment on 30-MAY-2021.

The total purchase value for this shipment is $ 128791.00

Regarding the payment and invoicing of Jeans it is correct that

(1) BIFT Fashion Group (Beijing) Co, Ltd pays the import duty and the freight charges.

(2) ANT DENIM TECHNOLGY LTD. covers all charges regarding exporting the goods.

BIFT Fashion Group (Beijing) Co, Ltd will be responsible for the customs duty declaration.

BIFT Fashion Group (Beijing) Co, Ltd will arrange the local transportation themselves.

It is confirmed that BIFT Fashion Group (Beijing) Co, Ltd will treat 17% Vat charged in connection with the import as incoming VAT.

Terms of payment: D/P AT SIGHT.

Terms of Delivery: FOB CHITTAGONG.

BIFT Fashion Group(Beijing) Co, Ltd.　　　　　　　　　　ANT Denim Technology Ltd.

　　　　　　　　　　　　　　　　　　　　　　　　　　For ANT Denim Technology Ltd.

　　　　　　　　　　　　　　　　　　　　　　　　　　————————————

　　　　　　　　　　　　　　　　　　　　　　　　　　Authorized Signature

Chapter 7

Consumer Demand and Fashion Marketing

课题名称：Consumer Demand and Fashion Marketing（消费者需求与时装营销）

课题时间：6课时

训练目的：通过本章的学习，学生可以了解经济因素对消费者需求的影响以及全球贸易对时装营销的影响；通过第二部分应用文的学习，了解在服装外贸中信用证的写法。

教学方式：教师主讲 Part A 的主课文和 Part B 的应用文，学生完成练习并课外阅读 Part A 后的 Supplementary Reading（补充阅读）。

教学要求：1. 使学生了解消费者需求与时装营销的相关知识。

2. 扩充学生服装方面的英文词汇。

3. 了解在服装外贸中信用证的写法。

Chapter 7
Part A

Consumer Demand and Fashion Marketing
消费者需求与时装营销

Fashion marketing is the entire process of research, planning, promoting and distributing the raw materials, apparel and accessories that consumers want to buy. It involves everyone in the fashion industry and occurs throughout the entire channel of distribution. Fashion marketing begins and ends with the consumer. It is the power behind the product development, production, distribution, retailing, and promotion of fibers, fabrics, leathers, furs, trimmings, apparel and accessories. [1]

Consumer Demand
消费者需求

The history of the fashion industry in America is the story of a growing economy, based on manufacturing, which consumed more than it could produce. As competition increased, consumers' demands have shifted the industry from a production to a marketing orientation. [2]

With this change in marketing philosophy, the industry now endeavors to find out what consumers will want to buy through research and then tries to develop the products to answer these needs. Fashion industry executives continually learn about consumer behavior to get clues as to what products consumers might need or want to buy in the future. They hire professional marketing firms to analyze life-styles and buying behaviors. Professional analysts have developed sophisticated marketing research methods to determine consumer wants and needs. In turn, manufacturers and retailers have emphasized product development to answer these needs.

Fashion producers and retailers also spend large amounts of money on increased advertising and other marketing activities to create consumer demand. The ultimate achievement of advertising is to establish the identity of a particular brand name or store so solidly that consumers will seek it out, that it will become a "destination brand". [3] Consumers are bombarded with million-dollar advertising campaigns. However, there is a limit to which marketing can win acceptance for a fashion. If the public is not ready for a product or is tired of it, no amount of advertising or publicity can gain or hold its acceptance.

The Impact of Economics on Consumer Demand
经济因素对消费者需求的影响

Consumer spending, the state of the economy, the international money market, and labor costs have an effect on fashion marketing.

Consumer Spending(消费者开支) The amount of money that consumers spend on fashion and other goods depends on their income. Income, as it affects spending, is measured in three ways: personal income, disposable income and discretionary income.

Purchasing Power(购买力) Purchasing power is related to the economic situation. Although incomes in the Western world have risen in recent years, so have prices. Thus, income is meaningful only in relation to the amount of goods and services that it can buy, or its purchasing power. *Credit, productivity, inflation, recession, labor costs, and international currency values affect purchasing power.*

The Impact of Global Trade on Marketing
全球贸易对时装营销的影响

A major trend in fashion marketing is globalization. World trade in apparel and accessories is growing despite high tariff and drastic currency fluctuations. Many countries may be involved in the production of a single garment.[4] For example, a garment could be designed in New York, made in China, and then distributed to retail stores all over the world. Retailers, too, are expanding globally. Stores as diverse as Christian Dior (French) and the Gap (American) have opened stores worldwide.

Imports(进口) Imports are goods that are brought in from a foreign country to sell here. The people or firms that import goods are usually manufacturers, wholesalers, or retailers, acting as importers. Retailers want to give their customers the best quality products at the lowest prices. *Therefore, ever-increasing amounts of textiles, apparel, and accessories are being imported into the United States and the European Union(EU) because of the availability of cheaper labor in low-wage countries. United States and European manufacturers compete with labor rates of 69 cents per hour in parts of China and 60 cents or less in India.*

Balance of Trade(贸易差额) The balance of trade is the difference in value between a country's exports and its imports. *Ideally, the two figures should be about equal. Lately, however, the United States has been importing much more than it exports sending American dollars abroad to pay for these goods and creating a huge trade deficit.* Many people believe that we should import fewer goods to balance trade. Others think that the consumer should be given the best merchandise at the best price, regardless of the balance of trade.

World Trade Organization(世界贸易组织) The World Trade Organization (WTO), which governs worldwide trade, is located in Geneva. The organization has members from 145 countries, who together account for over 90 percent of world merchandise trade. Its basic objectives are to achieve the expansion and progressive liberalization of world trade. *Additional functions of the WTO are to set rules governing trade behavior, set environmental and labor standards, protect intellectual property, resolve disputes between members, and serve as a forum for trade negotiations.* The Agreement on Textiles and Clothing(ATC) was negotiated under the auspices of the WTO. It provided for the 2005 quota removal from all apparel and textile imports for its 145 member countries.

Exports(出口) Because the US market is no longer growing and to help to balance imports, there is an increasing trend for American manufacturers to try to sell their merchandise to foreign retailers either through direct export or by licensing arrangements. However, import and export duties and a high dollar exchange rate have kept this business small.

The fastest growing economies in the world are the Southeast Asian countries, Eastern Europe and Latin America. At its current growth rate, China will have the largest economy in the world in 2025. No longer will these countries be simply a place to source goods, but they will also be attractive new markets because of their growing middle classes.

The Marketing Chain
服装营销链

The marketing chain is the flow of product development, production, and distribution from concept to consumer.

The traditional chain of marketing—textiles to apparel manufacturers to retailers to consumers—is no longer clearly divided. The old relationships between suppliers and retailers are disappearing and some new forms of marketing chain have come into being. [5]

Vertical Integration(纵向整合) Many companies are combining fabric production and apparel manufacturing or manufacturing and retailing, a strategy called vertical integration. [6] A completely vertical company produces fabrics, manufactures clothing, and sells the finished apparel in its own stores. Cutting out distribution costs (from manufacturer to retailer) increases profits and keeps prices down for the consumer. Vertical companies also like having total control of the supply, production and marketing chain.

Full Package Manufacturing(服装整体生产外包) With today's focus on global production, most American and European manufacturers and retailers design their own fashion, but buy "full garment packages" from contractors in Asia and elsewhere. *They source (find the best production available at the cheapest price) contract production to suppliers who can handle everything from buying piece goods to booking space on ships.*

Manufacturer-retailer Alliances(制造商—零售商联盟) Fierce competition has forced manufacturers and retailers to work together to achieve quick and cost-effective seamless distribution. They form informal partnerships or alliances to integrate the marketing chain. [7] Retailers discuss their needs with their manufacturer-partner; they work on product development together and plan production and shipping timetables together. For this arrangement to work, manufacturers and retailers must have complete trust in each other and open communications. In addition, many retailers have become manufacturers themselves.

Highlights

- Consumer demand has caused the fashion industry to convert from a production to a marketing focus.
- Economics and global trade have a great impact on both consumer demand and fashion marketing.
- The traditional marketing chain is no longer made up of completely separate levels. Fashion companies are expanding vertically or forming alliances to strengthen the chain.

For Review

1. Why has the fashion industry shifted from a production to a marketing orientation?
2. What impacts do the economics and global trade have on fashion marketing?
3. What are the characteristics of the new marketing chain?

New Words and Expressions

philosophy   *n.* 哲学,哲学体系;理念
bombard   *vt.* 炮轰;轰击
globalization   *n.* 全球化,全球性
tariff   *n.* 关税
currency   *n.* 货币
fluctuations   *n.* 波动
Geneva   *n.* 日内瓦城(瑞士西南部城市)
under the auspices of 由……赞助,由……资助
quota   *n.* 配额,限额
exchange rate   *n.* 汇率

executive   *n.* 执行者,经理主管人员
Intranet   *n.* 企业内部互联网
video(conferencing)   *n.* 电视会议
database   *n.* [计]数据库,资料库
vertical integration   *n.* [经]垂直统一管理,纵向整合
alliances   *n.* 联合
cost-effective   *adj.* 有成本效益的,划算的
seamless   *adj.* 无缝合线的,浑然一体的
partnerships   *n.* 合伙,合股;合伙企业

Translation

Translate the following sentences italicized in the text into Chinese.

1. Fashion marketing is the entire process of research, planning, promoting and distributing the raw materials, apparel and accessories that consumers want to buy.

2. Credit, productivity, inflation, recession, labor costs, and international currency values affect purchasing power.

3. Therefore, ever-increasing amounts of textiles, apparel, and accessories are being imported into the United States and the European Union (EU) because of the availability of cheaper labor in low-wage countries. United States and European manufacturers compete with labor rates of 69 cents per hour in parts of China and 60 cents or less in India.

4. Ideally, the two figures should be about equal. Lately, however, the United States has been importing much more than it exports sending American dollars abroad to pay for these goods and creating a huge trade deficit.

5. Additional functions of the WTO are to set rules governing trade behavior, set environmental and labor standards, protect intellectual property, resolve disputes between members, and serve as a forum for trade negotiations.

6. They source (find the best production available at the cheapest price) contract production to suppliers who can handle everything from buying piece goods to booking space on ships.

Websites

By accessing these Websites, you will be able to gain broader knowledge and up-to-date information on materials related to this chapter.

National Retail Federation: *http://www.nrf.com*

International Mass Retailers Association: *http://www.imra.org*

The Gallup Organization: *http://www.gallup.com*

US Department of Commerce: *http://www.commerce.gov*

AC Nielsen: *https://www.nielsen.com*

Notes to Part A

[1] Fashion marketing begins and ends with the consumer. It is the power behind the product development, production, distribution, retailing, and promotion of fibers, fabrics, leathers, furs, trimmings, apparel and accessories.

时装的市场营销始于消费者,终于消费者。它也是产品开发、生产、运送、零售的动力,并促进纤维、布料、皮革、毛皮、装饰、服装和配饰的生产。

[2] As competition increased, consumers' demands have shifted the industry

from a production to a marketing orientation.

随着竞争日益激烈,消费者需求使得服装业由以生产为主导转向了以市场营销为主导。

[3] The ultimate achievement of advertising is to establish the identity of a particular brand name or store so solidly that consumers will seek it out, that it will become a "destination brand".

广告宣传的最终目的就是使某个品牌或商店的形象深入人心,使消费者主动寻求它们,成为自己的"目标品牌"。

[4] A major trend in fashion marketing is globalization. World trade in apparel and accessories is growing despite high tariff and drastic currency fluctuations. Many countries may be involved in the production of a single garment.

时装市场营销的一个主要趋势就是全球化。尽管海关税收很高,外汇汇率起伏不定,但服装和配饰的海外贸易仍在增长。仅仅一件服装的生产就有可能会牵涉许多国家。

[5] The traditional chain of marketing—textiles to apparel manufacturers to retailers to consumers—is no longer clearly divided. The old relationships between suppliers and retailers are disappearing and some new forms of marketing chain have come into being.

传统的市场营销链是指从纺织品生产到服装生产,再从零售商到顾客的过程。但今天,这个过程之间的连接变得越来越模糊了。供应商和零售商之间以前的关系已经逐渐消失,出现了一些新形式的市场营销链。

[6] Many companies are combining fabric production and apparel manufacturing or manufacturing and retailing, a strategy called vertical integration.

很多公司开始把布料生产和服装生产结合起来,或者把服装生产和零售结合起来,这种策略被称为纵向整合。

[7] Fierce competition has forced manufacturers and retailers to work together to achieve quick and cost-effective seamless distribution. They form informal partnerships or alliances to integrate the marketing chain.

激烈的竞争迫使生产商和零售商展开合作,以达到快速、低成本的一体式配送。他们组成非正式伙伴关系或者联盟,将营销链整合在一起。

Supplementary Reading
Case study 1

The Disloyal Customer

Granvilles is a specialized department store located in the northeastern United States. It has been in operation for the past 50 years, operating first from its flagship store. With significant success, it opened 12 branch operations. Except for the flagship store, which is located in a busy downtown area, the other units are located in upscale, regional malls.

The store has always restricted its merchandise mix to men's, women's, and children's apparel and accessories. The price points are at levels that appeal primarily to upper middle class families. Until 3 years ago, Granvilles had been extremely profitable. Sales continued to increase, and management was satisfied with the showing.

During the past 3 years, however, there has been a noticeable decline in sales volume and profits. After carefully assessing its method of operation, management concluded that it was not doing anything different now than it had in the past. The only explanation for the decline in sales was competition from a major off-price retailer that was carrying some of the same designer labels at lower prices. Although Granvilles appealed to these manufacturers to stop shipping to its rival, the request was ignored. By selling to the off-price retailer, the manufacturers were able to dispose of leftovers(剩余物) at the season's end.

After numerous meetings, management decided that it could bring back some customers by adding private label merchandise to its inventory. Recognizing that the customer was important to the success of Granvilles, management decided to scientifically study the problem before making any final decision.

Questions

1. Has Granvilles properly defined its problem?
2. What type of research would you suggest that Granvilles undertake to solve its problem? How and why?

Case study 2

The Friendly Competitors

The downtown area of a major Midwestern city has been home to two large department stores for more than 50 years. Although each has a core of regular customers, they are in competition with each other. Both have fashion orientations, but Goldens is a little more fashion forward than the more traditional Baker & Foster. Like most fashion retailers, both organizations feature services for their shippers and initiate(发起) regular promotions and special events to increase customer traffic. They do not share company secrets with each other; however, their relationship has been amicable. Although several lines of merchandise are featured in both stores, they never seem to be involved in pricing disputes. Each works on the traditional retail markup(涨价,涨价幅度), reducing prices only when items fail to sell or at the conventional sales periods.

Yesterday's lead story in major local newspapers centered on the ground breaking for a new retail operation. Lamberts, a well-known off-price retailer announced that it will be opening a new unit 3 miles from the downtown area. Although the new business is not within walking distance of the old-line department stores, it is within easy reach via public and private transportation. Known for its shrewd merchandising practices, Lamberts features(以……为特色) well-known fashion merchandise at 20 to 50 percent below regular retail. Although Goldens and Baker & Foster receive their goods early in the season and Lamberts purchases later to gain a price advantage, Lamberts does pose a threat to the traditional retailers.

Management at both Goldens and Baker & Foster have called senior staff meetings to discuss plans for dealing with this potential new competitor. Several possible approaches have been suggested:

(1) Merchandise bearing the same labels should be discounted to meet the competition.

(2) New services should be offered to capture customers' attention.

(3) Service should be curtailed so that prices can be reduced throughout the store.

(4) Private labels should be increased.

(5) Lines carried by Lamberts should be discontinued.

Question

1. Describe the business characteristics of Goldens and Baker & Foster.
2. Which of the five suggested solutions could be employed to meet the challenge from Lamberts? Defend your reasoning.

Part B

Letter of Credit
信用证

信用证是一种由银行开立、有条件的书面付款承诺。在商业活动中,一般是开证行(Issuing Bank)根据开证申请人(Applicant)的请求和指示,向受益人(Beneficiary)开立的一定金额的、在一定期限内凭所规定单据(Documents)承诺付款的书面文件。信用证现已成为国际贸易中的一种主要付款方式,它把由买方履行付款责任转为由银行来履行付款,从而保证了卖方安全迅速地收到货款,买方按时收到货运单据。

信用证虽然没有统一的格式,但基本项目相同。主要包括以下项目:信用证号码;开证日期;信用担保金额;信用证议付条款;有关货物提单和货物的情况,主要说明货物名称、品质、规格、数量、单价等;有关运输的要求,主要有装运港和目的港、运输方式、装运期限、可否分批装运和可否转运;有关货物保险和赔付情况等。

卖方一般在收到信用证后方开始备货。若没有及时收到买方的信用证,可向买方发出信用证催促信。

Sample 1

以下是一个远期、不可撤销跟单信用证的实例。

Usance Irrevocable Documentary Letter of Credit No. 4551

From: Seattle Commercial Bank

Form of Documentary Credit: Irrevocable

Date and Place of Issue: February 15, 2021, Seattle, Washington, USA

Date and Place of Expiry: June 25, 2021, Shanghai, China

Applicant: Seattle Textile International Ltd., 16 Science Museum Road, Seattle, Washington, USA

Beneficiary: Shanghai Textile Import & Export Corporation, 44 Nan Jing Road, Shanghai, China Advising Bank: Bank of China, Shanghai Branch, 30 Nan Jing Road, Shanghai

Continued

Drawee: Issuing Bank

Amount: US Dollars 900000.00 (say US Dollars nine hundred thousand) Available with Seattle Commercial Bank by drafts at 30 days sight drawn on us for full invoice value of goods[1]

Partial shipment: not allowed

Transshipment: not allowed

Loading in charge: major China seaports

For transportation to: Seattle, Washington, USA

Latest date of shipment: March 15, 2021

Documents required:

Invoices in 4 copies;

Full set original, clean "on board" bills of lading, made out to the order of shipper, endorsed in blank, marked "freight prepaid" and "notify Seattle Textile International Ltd.".[2]

Marine insurance certificate for full CIF value plus 10% covering WPA and War Risk;

Packing lists in triplicate;

Country of origin in duplicate.

Additional condition: third party documents allowed and third party shipper allowed[3]

Shipment of goods: 10000 dozen Women Pajamas at $90 per dozen CIF Seattle, Washington USA

Presentation period: within 15 days after the issuance of the shipping documents but within the validity of the credit.

Instructions:

On receipt of documents conforming to the terms of this documentary credit, we undertake to reimburse you at maturity in the currency of this documentary credit in accordance with your instructions.[4] Except so far as otherwise expressly stated, this documentary credit is subject to uniform customs and practice for Documentary Credit (1993 version), International Chamber of Commerce Publication No. 500.

All documents must be dispatched to us by courier express delivery or by registered airmail in one lot.

New Words and Expressions

irrevocable   不可撤销的
documentary Letter of Credit   跟单信用证
issue   开证
expiry   终止,满期
endorsed in blank   空白背书
WPA = With Particular Average   单独海损,赔偿,担保单独海损,水渍险

applicant   申请人
beneficiary   受益人
advising Bank   通知行
drawee   付款人
country of origin   原产国,起运国
CIF = Cost Insurance and Freight   到岸价格
reimburse   偿还,偿付

Notes

[1] Available with Seattle Commercial Bank by drafts at 30 days sight drawn on us for full invoice value of goods.

议付行为西雅图商业银行,凭本证按发票金额100%开立以我行为付款人的30天远期信用证。

[2] Full set original, clean "on board" bills of lading, made out to the order of shipper, endorsed in blank, marked "freight prepaid" and "notify Seattle Textile International Ltd.";

全套已装船的正本清洁提单,空白背书并应注明:"运费预付""通知人:西雅图纺织品国际有限公司"。

[3] Third party documents allowed and third party shipper allowed.

接受联运提单。

[4] On receipt of documents conforming to the terms of this documentary credit, we undertake to reimburse you at maturity in the currency of this documentary credit in accordance with your instructions.

收到符合此跟单信用证条款的单据后,到期之后我方将按照你方指示对贵行进行偿付。

Sample 2

以下是一个催开信用证的实例(Urging to establish a L/C)。

Dear Sirs,

This is to refer to our Sales Confirmation No. 3645 covering 8000 dozen shirts for shipment on August 15th, 2021. We are disappointed to note that we have so far not received your L/C though you promised to establish it immediately after the signing of the contract. In the present circumstances, shipment would have to be postponed and certainly this is not our fault.

Please give this matter your immediate attention and take appropriate measure to fulfill your obligations without further delay.

Your prompt reply will be appreciated.

Yours faithfully,

(signed)

New Words and Expressions

Sales Confirmation 销售确认书

Chapter 8

The Fashion Retailer

课题名称：The Fashion Retailer（时装零售商）

课题时间：6课时

训练目的：通过本章的学习，学生可以了解时装零售商的分类、零售地点、零售商所提供的服务以及零售商如何树立形象；通过第二部分应用文的学习，了解在服装外贸中提单和索赔的写法。

教学方式：教师主讲 Part A 的主课文和 Part B 的应用文，学生完成练习并课外阅读 Part A 后的 Supplementary Reading（补充阅读）。

教学要求：1. 使学生了解时装零售的相关知识。

2. 扩充学生服装方面的英文词汇。

3. 了解在服装外贸中提单和索赔信函的写法。

Chapter 8
Part A

The Fashion Retailer
时装零售商

The hopes and dreams of fashion designers are ultimately in the hands of the fashion retailers. The design team can create an exciting new silhouette or coloration, but it is their retail counterparts who determine which products will be offered for sale.[1] Retail buyers and merchandisers screen all available merchandise before deciding which lines to display.

Classifying Retailers
零售商分类

Retail organizations in the 21st century bear little, if any, resemblance to the trading posts and general stores of earlier days. In the highly specialized and very competitive environments of today, retailers with fashion orientations cannot rely on the techniques of their predecessors.[2]

There are many types of retailers, each with a specific formula for attracting customers. Some are industrial giants; others, small entrepreneurs. The following discussion describes the different types of retail classifications, based on merchandise assortment and methods of operation.

Department Stores(百货商场) Department stores present a wide range of merchandise in defined areas or departments, selling both soft goods and hard goods. Consumers are mainly attracted by their fashion merchandise.

Usually department stores with strong fashion orientations are the most aggressive in terms of advertising and promotion. Macy's, Bloomingdale's and Marshall Field have almost daily fashion promotions in newspapers to capture the customer's attention.

Besides, department stores generally feature a category of apparel at many price points, each presented in a different area of the store. Thus, a store might sell sportswear in three different locations—main floor, third floor and sixth floor—but each concentrates on a separate price range or price point. Each area is assigned catchy names. [3] Bloomingdale's, for example, houses its moderate dress collection in Boulevard Dresses.

Specialty Stores(专卖店) *As merchandise became more varied, some early retailers moved away from the general merchandise concept and pioneered the first limited line stores, which concentrate on one merchandise classification.* Today, these specialty stores are a major force in retailing. And specialize they do!

This type of retail organization has two advantages. Because merchandise is restricted to one classification, the stores often feature the widest assortment available. In addition, shopping is faster and more convenient. With more people holding full-time jobs, the quick purchase has become a necessity.

Boutiques(精品店) A variation on the typical specialty store is the boutique. It is most often a one-unit operation that features upscale, fashion-forward merchandise. The assortment is generally restricted to just a few pieces of each item, with custom-tailored apparel sometimes featured. Customers frequent boutiques because they are usually guaranteed the latest in fashion innovation and are individually assisted by trained salespersons. [4]

Off-Price Retailers(折价店) Among the most successful retailers are the off-price merchants. Many prestigious labels are found in the off-price stores, like Jones New York, DKNY, Calvin Klein, etc. The off-pricers buy late in the season, when manufacturers are forced to close out their lines at reduced prices, and can sell the merchandise at reduced prices to consumers.

Flea Market Operations(跳蚤市场) All across the United States, shoppers are flocking to flea markets. These markets can be found in outdoor locations, such as movie parking lots, and in indoor facilities that once housed single retail operations. Many operate only on weekends; others, such as one of the nation's largest, The Swap Shop in Sunrise, Florida, are open 7 days a week.

Flea markets have many vendors, each with comparatively limited retail space and lower operating costs than traditional retailers. Vendors work on very low markups and are able to sell at considerably reduced prices.

Retailer Locations
零售地点

At the beginning of the 20th century, the only viable place to establish a retail operation was in a downtown area. Times have certainly changed! Malls of every size, shape, and image now serve the needs of consumers. Some are enclosed, others open; some are vertically constructed, others expanded horizontally.[5] Although downtown is still a vital retail location, retailing is no longer relegated to the traditional downtown shopping district.

Downtown Central Districts(市中心商业区) Most major department stores still operate from their downtown flagship stores. The executive headquarters for Macy's is in Herald Square, New York City, a downtown shopping area. *Not only are most merchandising and policy decisions made in these locations, but also the parent stores generally account for a significant part of the company's sales volume.*

Shopping Malls(购物中心) The first malls were built as outdoor shopping arenas in the 1950s, but today, the enclosed mall is the dominant type of retail location because it can offer a climate-controlled shopping environment.[6] The majority are horizontally constructed in suburban areas where land is plentiful and less expensive than in the downtown urban areas. Later, based on the success of suburban malls, developers have begun creating downtown malls in urban centers.

High-Fashion Centers(时尚购物街) Many major cities boast fashion centers that are not located in malls, congested downtown areas. These are generally shopping streets dotted with upscale fashion retailers. Their target markets are affluent consumers who seek the latest in both domestic and international styles, with price not a factor. These high-fashion centers include Madison Avenue in New York City, Worth Avenue in Palm Beach and Oak Street in Chicago.

Power Centers(小型购物场) Throughout the country, there are small shopping arenas known as power centers. They offer customers merchandise at highly discounted prices. The stores are usually very large fashion-oriented retailers, known for competitive pricing, and capable of drawing large crowds.

Services Offered by the Fashion Retailer
服装零售商提供的服务

The retail business is highly competitive; merchants must distinguish themselves from the competition in order to attract enough customers to turn a profit. One way to achieve this goal is by providing customer services. As a rule, those with price as the chief attraction offer the fewest services; those with traditional, fashion retailing operations, provide the most.

Personal Shopping(私人购物服务) Every upscale fashion emporium offers some kind of personal shopping. Most provide telephone assistance for shoppers who call the store with merchandise requests. Customers may also be invited into a salon to view and try on merchandise.

Corporate Purchasing(公司采购) Many upscale retailers have consultants who assist businesses in the purchase of presents for employees and clients. At peak gift-giving times, such as Christmas, this is an excellent program for generating increased sales. *The store wraps the gifts, like perfume, small leather goods, and home products, and sends them to designated clients with very little customer involvement.*

Interpreters(翻译服务) Tourists are a major source of revenue for retail establishments. To attract their business, retailers in the major cities employ interpreters to accommodate foreign-speaking visitors.

Gift Registries(礼物登记) To eliminate the purchase of unwanted or duplicate merchandise by gift givers, gift-receivers, such as newly-married couples or prospective parents, select the items they would like to receive and register their preferences with an appropriate store. Then, well-wishers may purchase one of these gifts on a visit to any branch store.[7]

Beauty Salon(美容沙龙) Many retailers include a beauty salon on the premises. Its purpose is to bring shoppers into the store. Once there, they might be motivated to make an unplanned purchase. Some fashion retailers provide informal modeling in the salons, so customers can view the range of the store's apparel and accessories.

Travel Services(旅游服务) Some retailers also operate leased travel departments. Although this service does not directly sell merchandise, it does provide an opportunity to familiarize potential travelers with the store's merchandise. As with beauty salons, these departments attract additional consumers to the store.

Restaurants(餐饮服务) Many retailers provide a variety of dining services from snack bars to elaborate restaurants. Dining facilities do more than feed hungry shoppers. They give shoppers an opportunity to relax within the store's environment. Afterward, the refreshed customer may be sufficiently motivated to continue to shop.

Gift Wrapping(礼物包装) A gift that is beautifully wrapped always makes a positive impression. Not only does the recipient feel special, but also he or she will remember the store when in need of a gift for someone else.

Alterations, delivery of merchandise, charge accounts, child care, and other services are provided by the retail industry. Each store must decide which services will generate enough business to warrant their inclusion, and which will help distinguish them from their competitors.[8]

Developing a Fashion Image
树立时尚形象

A retail organization's image can be determined from a review of its ads in newspapers and magazines. If every retailer had the same image, it would be difficult for the shopper to decide which one to patronize.

A major responsibility of the fashion-oriented retailer is to develop an image that will motivate shoppers to become customers. Those who successfully do this become the major players in the game of fashion.

Retailers advance their fashion images through promotion. Using a variety of techniques that include fashion shows, special celebrity appearances, fashion workshops, and visual presentations, the retailer tries to motivate the customer to come and see what all the excitement is about.

Highlights

- Traditionally, fashion retailers operate department and specialty store organizations, while an increasing number of retailers function as boutiques, off-pricers and flea market vendors.

- The most popular of the traditional shopping areas was the central downtown district. And the most preferred shopping environment is the enclosed shopping mall.
- To overcome increasing competition, retailers offer such services as personal shopping, corporate purchasing, language interpreters, gift registries and gift-wrapping.
- Retailers of apparel and accessories must develop fashion images that distinguish them from their competitors.

For Review

1. What are the classifications for fashion retailers?
2. Where are the typical locations for fashion retail outlets? What are the advantages and disadvantages of each?
3. In what ways do fashion retailers distinguish themselves from others?
4. How can fashion retailers attract as many customers as possible?

New Words and Expressions

counterpart *n.* 对应物,极相似的人或物;配对物
screen *v.* 筛选
resemblance *n.* 类同之处
orientation *n.* 方向,倾向性;定位;向东方
predecessor *n.* 前辈,前任;(被取代的)原有事物
formula *n.* 公式,规则
entrepreneurs *n.* 企业家
soft goods 纺织品,布匹等相关产品
hard goods (小汽车、电视机等)耐用品
feature *v.* 以……为特色
boulevard *n.* [美]林荫大道
pioneer *v.* 创办;打开(新领域)或准备(途径)
upscale *adj.* [美]高消费阶层的,(商品)质优价高的

custom-tailored *adj.* 定制的,定做的
viable *adj.* 可行的;能养活的;能生育的
relegate *v.* 转移;归入;提交
flagship store 旗舰店
sale volume 销售额
dominant *adj.* 有统治权的,占优势的,支配的
congested *adj.* 拥挤的
affluent *adj.* 富裕的,丰富的
emporium *n.* 商场,商业中心;大百货商店
premise *n.* [-s] 经营场所
leased *adj.* 租用的,租赁的
recipient *n.* 接受者
patronize *v.* 光顾,惠顾
celebrity *n.* 名人;名声
workshop *n.* 工作室;工厂,作坊

Translation

Translate the following sentences italicized in the text into Chinese.

1. Retail organizations in the 21st century bear little, if any, resemblance to the trading posts and general stores of earlier days.

2. As merchandise became more varied, some early retailers moved away from the general merchandise concept and pioneered the first limited line stores, which concentrate on one merchandise classification.

3. Flea markets have many vendors, each with comparatively limited retail space and lower operating costs than traditional retailers. Vendors work on very low markups and are able to sell at considerably reduced prices.

4. Not only are most merchandising and policy decisions made in these locations, but the parent stores generally account for a significant part of the company's sales volume.

5. The store wraps the gifts, like perfume, small leather goods, and home products, and sends them to designated clients with very little customer involvement.

6. A major responsibility of the fashion-oriented retailer is to develop an image that will motivate shoppers to become customers. Those who successfully do this become the major players in the game of fashion.

Websites

By accessing these Websites, you will be able to gain broader knowledge and up-to-date information on materials related to this chapter.

National Retail Federation：

http：//www.nrf.com

International Mass Retailers Association：

http：//www.imra.org

Notes to Part A

[1] The design team can create an exciting new silhouette or coloration, but it is their retail counterparts who determine which products will be offered for sale.

设计团队能够创造令人兴奋的新廓型或色彩,但是只有零售部才能决定销售哪种产品。

[2] In the highly specialized and very competitive environments of today, retailers with fashion orientations cannot rely on the techniques of their predecessors.

在高度专业化、竞争异常激烈的今天,销售定位为时装的零售商不能再依靠他们前辈使用的技巧了。

[3] Besides, department stores generally feature a category of apparel at many price points, each presented in a different area of the store. Thus, a store might sell sportswear in three different locations—main floor, third floor and sixth floor—but each concentrates on a separate price range or price point. Each area is assigned catchy names.

另外,一般来说,百货商场出售的一种类别的服装往往有不同的价位,在商场的不同地方销售。因此,商场里可能有三个不同的地方卖运动装———一层、三层和六层——但是每层的价位都不同,并且每个区域都有一个朗朗上口的名称。

[4] Customers frequent boutiques because they are usually guaranteed the latest in fashion innovation and are individually assisted by trained salespersons.

顾客之所以经常光顾那些时髦精品店,是因为他们在那能找到最流行的款式,而且能得到训练有素的导购员的专门服务。

[5] Malls of every size, shape, and image now serve the needs of consumers. Some are enclosed, others open; some are vertically constructed, others expanded horizontally.

现在,有各种不同规模、形式和形象的购物中心迎合了消费者多样的需求。有的是封闭式的,有的是露天式的;有的是垂直构建,有的是水平扩展。

[6] The first malls were built as outdoor shopping arenas in the 1950s, but today, the enclosed mall is the dominant type of retail location because it can offer a climate-controlled shopping environment.

第一批购物中心是在20世纪50年代建立的露天购物场,但今天,由于封闭式购物中心能够提供不受天气变化影响的购物环境,已经成为零售店铺的主导形式。

[7] To eliminate the purchase of unwanted or duplicate merchandise by gift givers, gift-receivers, such as newly-married couples or prospective parents, select the items they would like to receive and register their preferences with an appropriate store. Then, well-wishers may purchase one of these gifts on a visit to any branch store.

为了避免购买一些不需要的礼物或重复的礼物,礼物送出者和接受者,比如说新婚夫妇或准父母,会选好他们想要的东西,并在一个合适的商场登记。然后,祝福者会到这个商场的任何一个分店购买所登记的商品。

[8] Alterations, delivery of merchandise, charge accounts, child care, and other services are provided by the retail industry. Each store must decide which services will generate enough business to warrant their inclusion, and which will help

distinguish them from their competitors.

零售商还提供其他的一些服务,例如,送货、赊购账、照看儿童等。每个商场都必须判断哪些服务会为它们带来足够的能将这些服务费用包括在内的生意,哪种服务能使他们区别于其他的竞争对手。

Supplementary Reading
Case study 1

On-line Retailing

During the decade of the 1990s, words such as website, on-line, and Internet began to take center stage in the marketing of merchandise. Companies that previously restricted their selling to retail stores or catalogs have gone on-line. By combining traditional store-based sales with 24-hour Internet availability—the bricks-and-clicks concept—retailers are able to expand their reach to consumers, increasing customer convenience and in many cases providing a broader range of items for purchase than could be featured in their space-limited stores.

Traditional brick-and-mortar(实体的) chains are increasingly relying on the Internet to enable their customers to research items before driving to a store for purchase. A 1999 survey by the Internet research firm Jupiter Communications found that consumers spent more than $135 billion in stores and catalogs as a direct result of research they did on-line. In fact, 68 percent of shoppers contacted in a May 2000 survey reported that they had used the Internet to evaluate goods on-line before purchasing these items in a physical store.

With retailers expending significant money to develop their websites, and more and more fashion merchandise being offered on-line each day, the Internet is becoming an important retail outlet. Many industry watchers believe that the traditional brick-and-mortar chains that will fare best in this new environment are those that use the Internet to drive traffic between their stores and websites, or provide unique merchandise or discounts that are unavailable in their physical stores.

It is hard to predict the future of retailing in the 21st century.

Questions

1. What does "on-line retailing" mean?

2. Why do more and more retailers go on-line?

Case study 2

Computerized Kiosks

Interaction between consumers and sellers is becoming more and more popular. The idea was first used in nonfashion merchandising such as greeting cards, where individuals can tailor-make their messages. Today, the fashion world is adapting the concept to wearable products.

One of the early innovators in this concept, Levi-Strauss, launched a personal-fit program for woman in 1994. Based on the success of that program, the company introduced a more interactive(交互式的), unisex program under the name Original Spin. Consumers who use the program stand at a computerized kiosk and are taken through a menu of basic styling options such as classic, low-cut, and relaxed jeans from which they make their initial selection. Next, an assortment of colors such as authentic stonewash, light indigo(靛青), dark wash, black, or soft rigid is presented to the shopper. A variety of leg options is then presented; tapered(锥形的), straight, boot cut, wide or flair. Finally, the fly closure, zipper or button, is chosen. Once the preferences have been logged into the kiosk and a sales associate has taken the measurements, the information is electronically compiled and a test pair of jeans is tried on. Any number of test pairs may be tried until complete satisfaction is achieved. Once all of the adjustments have been made, the order is sent to a Levi's factory in Tennessee(田纳西州) where the jeans are constructed. The order is promised to the customer within 15 days. For purposes of identification, and possible future orders, each front pocket of the selected jeans is finished with a bar code that indicates the customer's personal made-to-measure information. If adjustments are necessary for subsequent orders, such as leg shape or color, they may be called in to the company. The cost of this customized product is $55, and a full refund is promised if the purchaser is not completely satisfied.

Through this type of interaction, the shopper is involved in the product's design, something that often adds to satisfaction. Although the computerized kiosk concept is in its infancy, more companies are looking into it for future use with other fashion merchandise.

Questions

1. What are the advantages of the computerized kiosk?
2. Does this form of retail purchasing appeal to you?

Part B

Bill of Lading & Claims
提单和索赔

提单,是指用以证明海上货物运输合同和货物已经由承运人接收或者装船,以及承运人保证据以交付货物的单证。提单中载明的向记名人交付货物,或者按照指示人的指示交付货物,或者向提单持有人交付货物的条款,构成承运人据以交付货物的保证。

提单内容,包括下列各项:货物的品名、标志、包数或者件数、重量或者体积以及运输危险货物时对危险性质的说明;承运人的名称和主营业所;船舶名称;托运人的名称;收货人的名称;装货港和在装货港接收货物的日期;卸货港;多式联运提单增列接收货物地点和交付货物地点;提单的签发日期、地点和份数;运费的支付;承运人或者其代表的签字。

提单缺少前款规定的一项或者几项的,不影响提单的性质。并且,承运人或者代其签发提单的人未在提单上批注货物表面状况的,视为货物的表面状况良好。

Sample 1

以下是一个提单的实例。

Bill of Lading (B/L)

China Ocean Shipping Company

Shipper: Mr. Li Ming	head Office: Beijing
Consignee: Mr. Park Davis	Branch Office: Shanghai
	Cable Address: "COSCO"
Notify New York, USA	Bill of Lading
	Direct or with Transshipment
Vessel: Chang Hong Voy.	S/O No.: 001874
	B/L No.: 001412
Port of Loading: Shanghai	Port of Discharge: New York
Nationality: P. R. China	Freight Payable at: Shanghai, China

Particulars Furnished by the Shipper

Description of Goods	Marks and Numbers	No. of Packages	Weight		Measurement
			Gross	Net	
Pajamas	CNC 119	450Cartons	5840kg	4940kg	70 cm × 37 cm × 23 cm Dimension: 27CU. M.
Total Packages (in words): ONE HUNDRED CARTONS					

New Words and Expressions

shipper　托运人　　　　　　　　　port of loading　装货港
consignee　收货人　　　　　　　　port of discharge　卸货港
cable address　电报挂号　　　　　mark　标志
direct　直运　　　　　　　　　　description of goods　货名
transshipment　转运,转船　　　　CU. M.　立方米
vessel　船　　　　　　　　　　　in words　用文字表明
S/O(shipping order)　装货单

　　在履约过程中,尽管有关各方工作细致,但还是会发生麻烦,比如货物损坏、错发,未按时抵达,短重,质量不合格等。按合同规定,受损方有权向造成损失的一方提出索赔。

　　一封索赔信函不应过度指责他人,而应以简要的形式提供全部细节,友好、有理、有利、有节地提出索赔。这类信函可遵循下列步骤:

　　(1)开首对提出抱怨表示遗憾,并提及订单日期及编号、产品型号或目录、交货日期及所抱怨的货物名称乃至价格。

　　(2)陈述不满的原因,并通过描述差错的具体情况说明给你带来的不便和蒙受的损失,要求对方做出解释。

　　(3)提出索赔,陈述自己的理赔意见,并盼望对方早日理赔。

Sample 2

以下是对短重和质量不合格的索赔实例。

August 16, 2021

Dear Sirs,

　　With reference to our telephone of August 16, in connection with the down jacket shipped per S. S "Maria" in execution of contract No. AC-5180, we provide the following information in detail. A thorough examination by commodity inspection organization concerned showed that the short weight was due to the improper packing, the inferior quality was due to the deficiency of down content. We now lodge claims with you as follows:

Claim Number	Claim for	Amount
BPC 87	short-weight	US $ 1354. 30
BPC 88	Inferior quality	3201. 14
	Plus survey charges	68. 25
	Total Amount	4623. 69

　　To support our claims, we enclose one copy of the Survey Report together with our Statement of Claim which amounts to US $4623. 69.

　　We feel sure that you will give our claims your most favorable consideration and let us have your settlement at an early date.

Yours sincerely,

……

New Words and Expressions

down jacket　羽绒夹克　　　　　　　packing　包装
per　经,由　　　　　　　　　　　　inferior quality　质量次的
S. S x steam ship　蒸汽轮船　　　　deficiency　缺乏,不足
in execution of　执行　　　　　　　lodge　提出
inspection　检查　　　　　　　　　favorable　赞成的;有利的;起促进作用的

　　无论对方索赔理由是否成立,答复索赔的信函均应彬彬有礼。书写理赔(Adjustment of Claim)信函需遵循以下原则:

　　(1)首先应确定对方的抱怨和索赔是否成立,若事实存在,要欣然承认,表示歉意,答应赔偿。

　　(2)若对方的索赔要求不成立,就要礼貌地以可接受的方式指出来,详细说明原因和情况。

　　(3)如若不能立即做出答复,告诉对方问题正在调查中,一旦事情澄清,将

立即予以答复。

Sample 3

以下是对上封索赔信的理赔实例。

August 27, 2020

Dear Sirs,

Re: Contract No. AC-5180-Down-Jacket

We acknowledge receipt of your letter of August 16, with enclosures, claiming for shortage in weight and inferior quality on the consignment of Down Jacket shipped by S. S. "Maria".

Having immediately looked into the matter, we find that our down jacket was properly weighed at the time of loading and the quality of the consignment was up to standard. We really cannot account for the reason for your complaint. Since the goods were examined by a public surveyor, we cannot but accept your claims as tended.

We therefore enclose our check No. 44312 for US $4623.69 in full and final settlement of your claims BPC87 and 88. Kindly acknowledge receipt at your convenience.

We believe this is a fair adjustment of your claim, and trust that it will be completely acceptable to you.

Yours truly,

(Signed)

New Words and Expressions

with enclosure　带有附件
consignment　交货
look into　调查
be up to　达到
surveyor　检察员

tend　提出;注意;管理
in full　用完整的词(不缩写),以全文
settlement　解决
acknowledge　承认,确认
at one's convenience　尽早

Chapter 9

Advertising, Special Events and Visual Merchandising

课题名称：Advertising, Special Events and Visual Merchandising（广告、专门促销活动和视觉销售）

课题时间：6课时

训练目的：通过本章的学习，学生可以了解时装广告、专门促销活动和视觉销售的重要性和方式；通过第二部分应用文的学习，了解在服装行业求职时个人简历的写法。

教学方式：教师主讲 Part A 的主课文和 Part B 的应用文，学生完成练习并课外阅读 Part A 后的 Supplementary Reading（补充阅读）。

教学要求：1. 使学生了解时装广告、专门促销活动和视觉销售的相关知识。

2. 扩充学生服装方面的英文词汇。

3. 了解在服装行业求职时个人简历的写法。

Chapter 9
Part A

Advertising, Special Events and Visual Merchandising
广告、专门促销活动和视觉销售

Everyone in fashion recognizes the importance of effective promotion. *Designers, manufacturers, and retailers of fashion merchandise pay as much attention to promoting their products as they do to designing and merchandising them.* Elaborate runway shows, video presentations, and multimedia advertising campaigns are just some of the methods that can be used to introduce new collections, seasons, styles, and designs to an eagerly awaiting audience.[1] The cost of these promotional undertakings sometimes runs into the millions. Calvin Klein, for example, spent $40 million to launch a new fragrance.

Advertising
广告

"That paid for form of nonpersonal presentation of the facts about goods, services or ideas to a group" is the American Marketing Association's definition of advertising. Traditionally advertisers have used both broadcast and print media to get their messages across. Today, electronic advertising is also being carried over the Internet. In the fashion industry, advertising sponsors include trade organizations, designers, manufacturers and retailers. Each sponsor attempts to address the target market for his or her products.

Designers(设计师) Many designers have advertising needs that are two fold in nature. *One is to make the trade aware of their creations—that is, to capture the attention of the store buyers and merchandisers who make selections for their particular clienteles.* The other is to reach the ultimate consumer. Thus, designers advertise in particular trade periodicals to motivate store buyers and in consumer magazines and on television to appeal to ultimate users.

Manufacturers(生产商) The producers of fashion merchandise spend significant sums on advertising. Their targets are generally the retailers who are their potential customers. Advertising of this nature usually appears in trade papers and magazines or through direct marketing. Many manufacturers secure the names of potential accounts from marketing research organizations and then mail brochures, flyers, and videos that depict their offerings to them.

Retailers(零售商) *The major share of the retailer's promotional budget is earmarked for advertising; the major portion of the advertising budget is spent on newspaper ads.* Many of the major retailers have large staffs that are responsible for advertising and also have invested in computer hardware and desktop publishing software, so they can create their own catalogs. [2] This has cut production expenses considerably in this area of advertising.

Advertising Classifications(广告分类) Two distinct types of advertisements are used to gain customer attention. *Whether the ads are trade oriented and directed toward the industry or consumer based and focused on the ultimate consumer, the formats are either promotional or institutional.*

Promotional advertising, or product advertising as it is sometimes called, is used to sell specific items. Institutional advertising, on the other hand, directs its efforts toward projecting a particular image, achieving goodwill, or announcing special events. Advertisers may choose one of them according to their purposes of advertising, but sometimes they use a combination of both approaches.

Promotional advertising far outweighs the use of institutional advertising. Its positive effect can be quickly measured by increases in sales for the advertised items. Institutional results cannot be measured as scientifically or as quickly. Because the store's image is the focus of such ads, success can only be judged over a long period of time.

Special Events
专门促销活动

Retailers, manufacturers, designers, and trade associations each have many avenues for reaching both regular audiences and potential users of their products or services. [3] These special events may be major attractions that cost significant amounts and last for several weeks, or they may be less costly one-day affairs. Fashion shows, celebrity appearances, theme parades, demonstrations, charitable celebrations, and special sales are just some of the events in which the fashion industry participates.

Fashion Shows(时装表演) Few special presentations offer the drama and excitement of fashion shows. Whether the audience is composed of professional industrial purchasers or consumers, the live production seems to excite everyone. Budgeting considerations, space, audience size, and purpose play a role in deciding the format of a fashion show. Once these factors have been addressed, the production will follow one of two forms.

The runway show is the most elaborate type of fashion show. These shows require music, either live or recorded, choreography, scripts and models. Informal modeling is the second format used. As the name implied, models walk among customers showing off selected outfits. In store restaurants and beauty salons, models parade the latest fashions in the hope that consumers will be motivated to buy the merchandise.

Personal Appearance(名人出场) One of the surest ways to bring an audience to a store is to advertise a personal appearance by a celebrity. Each industry, including the fashion industry, has charismatic personalities whose very presence will guarantee large crowds. Whether it is Calvin Klein or Karl Lagerfeld promoting a fragrance, Betsey Johnson talking about her newest collection, or Anna Sui showing her latest line, the results are usually successful.

Demonstrations(产品演示) Capturing the shopper's attention sometimes necessitates a demonstration of how a product may be used. The cosmetic industry often uses the demonstration technique to entice customers to purchase its products.

Mobile Presentations(流动展示) Unlike the typical designer special events, which require attendees to come to a company's own venue or a space that it has rented for the occasion, this new format enables the event or promotion to move from one place to another in a mobile vehicle. [4] These "traveling road shows" are akin to libraries on wheels or medical clinics that look to disseminate information.

Sampling(样品试用) This type of promotion requires giving away products to prospective users. In the retail stores where their products are sold, the manufacturers provide sample items or kits that are given free with a purchase or provided at a minimal cost.

Premiums(额外奖励) *Sometimes fragrance and cosmetic vendors reward purchasers of their products with premiums that are free or comparatively inexpensive.* The items include umbrellas, luggage, carrying cases, T-shirts, and other items that generally bear the vendor name whenever the premium is used.

Personal Improvement Sessions(个性时尚咨询) Many fashion retail operations present seminars on personal grooming and proper dress. These events are usually held in a store's special events center or community room and features a fashion consultant who discusses the "do's and don'ts" of appropriate dress.

Visual Merchandising
视觉销售

Upon arrival at a retail operation or a manufacturer's showroom, customers should be greeted with an environment carefully designed to further arouse their interest. Designed to capture on-premised attention, visual presentations enhance a company's selling and display areas by establishing a climate in which sales will be made. This is an integral part of the areas of promotion already explored.

The design and execution of visual presentations may be coordinated by full-time company teams headed by visual merchandising directors, many of whom carry the title of vice-president, an indication of the position's importance.[5] *The presentation can also be produced by consultants, who are paid by companies to design settings and execute display installations, and by freelancers, who periodically install visual presentations.*

In retailing, visual merchandising is generally broken down into two areas—windows and interiors.

Window Displays(橱窗展示) The silent sellers for many retailers are the windows that line the streets and malls. In the downtown flagships, careful attention is paid to window displays. Usually changed once a week, the themes might include a holiday such as Christmas, a special salute to a designer's new collection, a specific sale period, the introduction of a store's new private label, or anything that might attract the attention of passersby. These visual stories range from the unique, such as the animated presentations at Christmas time, to the traditional. Whatever the event or occasion, they must be executed with props and lighting that enhance the display.

Interior Displays(室内展示) Inside the stores, visual merchandisers are regularly installing displays, and improving the general appearance of the store. With the enormous cost of retail rentals, many merchants are reducing the amount of space for formal windows, then, interior displays are given greater attention.[6]

One of the trends in interiors is the elimination of props that depict particular seasons or holidays. For example, Gap merely changes its merchandise displays in the store, without using any holiday or seasonal symbols.

Highlights

- Fashion organizations of every type and size develop a variety of promotional programs to enhance their images and sell merchandise, including advertising, special events and visual merchandising.
- Manufacturers, designers, and retailers use all forms of media to advertise.
- Special events are periodic presentations, such as fashion shows, personal appearances, product demonstrations, mobile presentations, product sampling, premium offers, and personal improvement sessions.
- Visual merchandising is used by retailers to give their showrooms more eye appeal to motivate prospective purchasers.

For Review

1. Describe different needs of advertising sponsors.
2. Differentiate between promotional and institutional advertising.
3. What are the two formats for fashion shows?
4. In what way does visual merchandising play an important role in product promotion? What are two types of visual display?

New Words and Expressions

multimedia *n.* 多媒体
undertaking *n.* 事业,企业;承诺
electronic *adj.* 电子的
sponsor *n./v.* 发起人,赞助人,主办人;赞助
twofold *adj.* 两部分的,双重的
clientele *n.* 客户
charitable *adj.* 仁慈的,(为)慈善事业的

Chapter 9 Advertising, Special Events and Visual Merchandising

choreography　*n.*（芭蕾舞或诗歌等）舞步设计与编排艺术
charismatic　*adj.* 神赐能力的；超凡魅力的
personality　*n.* 个性，人格；人物，名人
entice　*v.* 诱惑，诱使
guarantee　*v.* 保证，担保
be akin to 类似
periodical　*n./adj.* 期刊，杂志；周期的，定期的
potential　*adj.* 潜在的，可能的
flyer　*n.*（广告）传单
budget　*n./v.* 预算；做预算
earmark　*v.* 指定（款项等的）用途
institutional　*adj.* 制度上的
kit　*n.* 成套工具，用具包，工具箱，成套用具
vendor　*n.* 卖东西的小贩；卖主
seminar　*n.* 专题讨论课；短期强化课程
installation　*n.* 安装；装置
freelancer　*n.* 自由职业者
animated　*adj.* 动画的，栩栩如生的
prop　*n.* 服装等道具

Translation

Translate the following sentences italicized in the text into Chinese.

1. Designers, manufacturers, and retailers of fashion merchandise pay as much attention to promoting their products as they do to designing and merchandising them.

2. One is to make the trade aware of their creations—that is, to capture the attention of the store buyers and merchandisers who make selections for their particular clienteles.

3. The major share of the retailer's promotional budget is earmarked for advertising; the major portion of the advertising budget is spent on newspaper ads.

4. Whether the ads are trade oriented and directed toward the industry or consumer based and focused on the ultimate consumer, the formats are either promotional or institutional.

5. Sometimes fragrance and cosmetic vendors reward purchasers of their products with premiums that are free or comparatively inexpensive.

6. The presentation can also be produced by consultants, who are paid by companies to design settings and execute display installations, and by freelancers, who periodically install visual presentations.

Websites

By accessing these Websites, you will be able to gain broader knowledge and

up-to-date information on materials related to this chapter.

Advertising Agencies:

http://www.advertisingagencies.org/

Notes to Part A

[1] Elaborate runway shows, video presentations, and multimedia advertising campaigns are just some of the methods that can be used to introduce new collections, seasons, styles, and designs to an eagerly awaiting audience.

精心设计的时装表演、宣传录像片和多媒体广告只是向翘首企盼的观众推出新的服装系列、新的季节性服装和新设计的几种方式。

[2] Many of the major retailers have large staffs that are responsible for advertising and also have invested in computer hardware and desktop publishing software, so they can create their own catalogs.

许多大的零售商都拥有自己庞大的广告团队,同时,还投资于计算机硬件和桌面出版软件来创作他们自己的产品目录册。

[3] Retailers, manufacturers, designers, and trade associations each have many avenues for reaching both regular audiences and potential users of their products or services.

零售商、生产商、设计师和贸易协会都有许多途径,用以联系使用他们的产品或服务的老顾客和潜在顾客。

[4] Unlike the typical designer special events, which require attendees to come to a company's own venue or a space that it has rented for the occasion, this new format enables the event or promotion to move from one place to another in a mobile vehicle.

不像一些典型的设计师推广活动那样,要求参加者必须到公司自己的或租用的展示现场,这种新方法使推广活动在一个可移动的车辆中进行,可以从一个地方移到另一个地方。

[5] The design and execution of visual presentations may be coordinated by full-time company teams headed by visual merchandising directors, many of whom carry the title of vice-president, an indication of the position's importance.

视觉展示的设计和制作可以通过一个由视觉销售主管领导的全职团队来协调进行,这些主管中有些还挂着公司副总裁的头衔,由此表明该职位的重要性。

[6] With the enormous cost of retail rentals, many merchants are reducing the

amount of space for formal windows, then, interior displays are given greater attention.

由于零售租金的巨额花费,许多店主都在减少正规的橱窗面积,因此,他们对室内的产品展示就更加关注。

Supplementary Reading
Case study 1

The Case of the Cost-Free Advertising Campaign

Major manufacturers and retailers set aside large sums of money promoting their merchandise and their companies. They recognize the value of such investments and make certain that their budgets are sufficient to reach potential customers.

Barbara Simms fully understands the need for promotion. In college, she learned all about the benefits, and she witnessed, firsthand, the returns realized from such activities when she worked for a major department store. The store, Atlees, Ltd., spent a great deal of money on advertising and extravagant(奢侈的) special events. It reaped extra benefits from the publicity derived from the special presentations. The company not only invested heavily, but it also had a large in-house staff that could create professional ads, build props, and create exciting promotional themes.

Recently, Barbara left Atlees, Ltd. and opened a small neighborhood fashion boutique. The initial costs of opening the shop were more than she anticipated and little was left for promotion. She would like to spend her limited resources wisely for a good advertisement and, also, present a cost-free special event that would make her trading area aware of her boutique(服装店). Her problem is how to coordinate an ad and special event without straining her budget.

Questions

1. How would you design a special event that would be virtually cost-free?
2. Can you devise an advertisement that would be inexpensive and compatible with the event for Barbara?

Case study 2

Macy's

One of the reasons for the enormous interest in Macy's is the wealth of promotions it produces. As a leading world fashion retailer, its fashion shows range from introducing a new designer collection to charity benefits. The company, however, does not stop there. It is Macy's institutional promotional events that separate the store from most others.

Macy's Fourth of July fireworks display is an exciting event that is viewed by thousands of people from New York City's waterfront (水边码头区,滨水地区) or on board local boats. Eleven thousand display shells and effects are exploded to create more than a million bursts of color and light, playing against a specifically written musical score. It is syndicated on television to 150 stations. This production brings Macy's a great deal of publicity.

For 2 weeks every spring, the Flower Show has become another Macy's tradition. Its Herald Square flagship main floor of 265000 square feet is filled with the exotic plants. A major spectacle, it is planned more than 1 year in advance. Floral experts throughout the world collect their finest specimens for showing. An additional feature consists of home fashion displays designed by celebrities.

Tap-O-Mania is an event that grows in size every year. For one Sunday each August, a time when shoppers are not akin to enthusiastically filling the store, Macy's puts on the largest tap dancing festival in the world. Participants, with or without tapping experience, are invited to join in the fun. Each year, the event attracts more than 6000 tap dancers, who perform in front of the flagship's entrance, attracting scores of shoppers. Unlike the other promotions, which cost large sums of money, this one is relatively inexpensive.

Through such unusual undertakings(事业), Macy's has established itself as the country's premier(第一的) retail promoter.

Questions

1. How has Macy's established itself as the country's premier retail promoter?
2. What have you learned from this case study?

Chapter 9　Advertising, Special Events and Visual Merchandising

Part B

Résumé
(个人简历)

个人简历是求职者生活、学习、工作、经历、成绩的概括。一份适合职位要求、翔实的简历可以有效地获得聘用单位面试的机会。在书写英文简历时，应注意如下几个原则：

(1)英文简历没有固定的格式，需要自行设计，应打造最能凸显自己优势的内容呈现方式。

(2)简历应以条列编排为原则，让主试者在短时间内就能抓住重点。

(3)简历尽量控制在一页纸之内，版面要简洁大方。

(4)英文简历前一定要搭配求职信(Cover Letter)，表明自己写信的原因和目的，简述自己的工作、教育背景和特长，提供备询人员或推荐人，结尾请求面谈的机会等。

(5)英文简历并不需要附上照片，除非公司有所要求。

Sample 1

以下是一封应聘助理设计师(Assistant Designer)的求职信。

Dear Sir/Madam,

　　I am very interested in your advertisement for an assistant designer in the October 28, 2020 in Beijing Youth Weekly because the position you described sounds exactly like the kind of job I am seeking.

　　According to the advertisement, your position requires graduates of Bachelor or above in fashion design, and I feel that I am competent to meet the requirements. I will graduate from Beijing Institute of Fashion Technology this year with a M. S. degree. My studies have included courses in fashion design, product development, fashion management and other related ones. During my education, I have grasped the principal of my major and skills of practice.

　　As for my English level, I've not only passed CET-6, but also more important I can communicate with others freely in English especially with the foreigners.

　　I would appreciate your time in reviewing my enclosed resume and if there is any additional information you require, please contact me. I am looking forward to an opportunity to meet with you for a personal interview. My mobile phone number is 13101234567.

　　With many thanks.

Sincerely,

Wang Lin

Sample 2

以下是一封应聘助理设计师的个人简历。

Room 212 Building 343
Beijing Institute of Fashion Technology, Beijing 100029
(010)62411546 Email:zhengyu@bift.edu.cn

Zheng Yu

Objective

To obtain a challenging position as an assistant designer

Education

2017. 9-2020. 6 Dept. of Fashion Design and Engineering of Beijing Institute of Fashion Technology, M. E.

2013. 9-2017. 7 Dept. of Fashion Design and Engineering of Beijing Institute of Fashion Technology, B. E.

Academic Main Courses

Fashion Design, Product Development, Fashion Management, Marketing for clothing, Psychology of Dress, Analysis of Consumer Behavior for Fashion, Pattern Making, etc.

Computer Abilities

Skilled in use of CAD software, Office 2016, Windows 10

English Skills

Have a good command of both spoken and written English

Past CET-6, TOEFL:602

Scholarships and Awards

- 2020. 3 First-class Scholarship for graduate
- 2016. 4 Academic Progress Award

Others:

Aggressive, independent and be able to work under a dynamic environment. Have coordination skills, teamwork spirit. Diligence and dedication are my greatest strengths.

Portfolio available upon request

New Words and Expressions

competent   有能力的,胜任的   enclosed   被附上的
principal   主要的,首要的

Chapter 10

Careers in Fashion

课题名称：Careers in Fashion（时装产业的职业机会）

课题时间：6课时

训练目的：通过本章的学习，学生可以了解服装纺织业、服装生产业、服装零售业和服装传媒中的各种职位及职能；同时通过第二部分应用文的学习，了解在服装行业中招聘广告的写法。

教学方式：教师主讲 Part A 的主课文和 Part B 的应用文，学生完成练习并课外阅读 Part A 后的 Supplementary Reading（补充阅读）。

教学要求：1. 使学生了解服装业中各种职位的职能。

2. 扩充学生服装方面的英文词汇。

3. 了解在服装行业中招聘广告的写法。

Chapter 10
Part A

Careers in Fashion
时装产业的职业机会

The world of fashion offers challenging, exciting, and financially rewarding career opportunities. The variety of activities involved in fashion results in diverse jobs that attract people with different backgrounds. Unlike other industries and professions, rigorous credentials and licenses are not required. *Although formal education and training are beneficial, successful people in fashion may have studied fine arts, marketing, design, textiles, or just a broad-based program.* Many legends began their careers in other fields before choosing fashion. Giorgio Armani studied medicine, as did Geoffrey Beene; Gianni Versace and Gianfranco Ferre studied architecture; and Vivienne Westwood and Bruce Oldfield were teachers.

This chapter discusses the fashion industry in terms of the careers offered in textiles, manufacturing, retailing, market consultants and fashion communications.

Textiles
纺织领域

The textile industry employs technically skilled individuals, whose talents and abilities range from artistic creativity to production and sales.

Textile Designer(纺织品设计师) Textile designers are artists who create particular patterns and present them in a format that can be translated into fabrics. They paint their designs on paper or fabric and prepare the repeats that will be used in the finished products.[1] Their extensive use of computers necessitates a thorough understanding of the available software. Some major companies employ designers who simply develop the design concepts and leave the technical developments to repeat artists and painters. Those who usually enter this aspect of the industry are art and design graduates. Their remuneration is generally high.

Colorist(色彩设计师、配色师) In companies with a great deal of specialization, the colorist is responsible for creating the color combinations that will be used in the production of the designer's patterns. The colorist must be an expert in color theory and must understand all of the technical aspects of color utilization.

Grapher(制图员) In knitwear, after an initial design has been completed, the design is graphed. These graphs are then used in the production process. In addition to requiring a complete knowledge of knitting construction, the grapher must also be computer literate. Computer-aided design (CAD) programs now allow graphing to be accomplished more quickly.

Converter(布料加工批发商) The converter oversees the change of gray goods, which are unfinished fabrics, into finished textiles. *Dyeing, printing, and the application of a variety of fabric finishes constitute converting.* Some of the finishes enhance appearance, whereas others are merely functional. The converter's career is a highly technical one that requires a complete knowledge of fibers and fabrics.

Dyer(染色师) A comprehensive knowledge of dyeing techniques, dye substances, colors, and chemicals is the responsibility of the person who dyes the stock, yarns, or finished fabric. The dyer is actually a textile chemist who understands all of the interactions of fabrics and the colors that will be applied. He or she should be a graduate of a textile chemistry program to ably perform the tasks involved in dyeing.

Textile Sales Representatives (Reps)(纺织品销售代表) Designers, manufacturers, and retailers are customers of textile companies. Whatever the market they serve, sales reps in textile companies must have a complete understanding of fibers, weaving and knitting, coloring and finishing processes, product care, and fabric end uses. Knowledge is acquired through both formal and on-the-job training. These professionals have the potential to earn substantial incomes.

Manufacturing
服装生产领域

At the very core of the fashion industry is the manufacturer. Whatever the products, manufacturing positions cut across all lines.[2]

Designer(设计师) *Designers are responsible for setting the tone of a line in* *terms of silhouette*, *color*, *fabrication and trim*. The most successful have an educational background that includes sketching, draping, pattern making and sewing. They are the mainstays of the industry – without their creations, there would be no lines to sell.

Designers are so vital to the company; they are often the principals in the business and have their names on the labels. The major fashion houses are known by the names of their designers; they are equivalent to stars of the entertainment industry.

Merchandiser（贸易商）The responsibilities of this position vary from company to company, but generally include making decisions concerning the company's line and fabric, marketing research, projecting sales, serving as liaison with the sales staff, contacting the mills, reviewing production considerations, and costing the merchandise. Merchandisers are generally highly paid, especially when product development is within their control.

Stylist（风格师、服装造型师）Companies that do not have a designer or do not give the merchandiser total responsibility for style development may employ a stylist. Stylists travel extensively, visiting markets to select styles that will fit the company's line. Rather than just copying the originals, the stylist generally translates each style to fit the manufacturer's needs. A thorough knowledge of textiles and color is necessary, so that the stylist can substitute fabrics and colors in the original designs to make the copy cost-efficient. [3]

Patternmaker/Grader（打板师/推板师）Patternmakers use the original design to create the patterns that will be used to produce the finished garments. They must be technically trained in construction, grading of sizes, production, cutting, and fabric utilization. Because most companies now use computers for patternmaking and grading procedures, patternmakers and graders need a working knowledge of computer programs and can use a digitizer for grading patterns. Salaries for these positions are very high because few people choose to specialize in this area.

Cutter（裁剪师）As the name implies, cutters cut the fabrics and other materials into shapes as dictated by the patterns. This career requires considerable technical skill, including familiarity with computerized cutting. Though the skill is very important, remuneration is moderate.

Quality Controller（质量控制员）One problem that often plagues the manufacturer is poor quality. Many companies employ quality controllers to make certain that the merchandise headed for the retailer is in the best condition to guarantee customer satisfaction. Quality controllers are well paid because their expertise enables the manufacturer to produce the best possible merchandise.

Retailing
服装零售领域

Retailing is a field that offers career opportunities at many levels and in many locations. The followings are some of the important careers in fashion retailing.

General Merchandise Manager (GMM.) (销售总经理) General merchandise managers supervise the store's several divisional merchandise managers. *They allocate dollars to be spent on each division's merchandise offerings and instruct the divisional merchandise managers about the store's markup and profitability goals, image, and other policies with a merchandising orientation.* GMM. s are among the highest paid retail executives.

Buyer (买手) Buying is seen as the glamour career in the store. Although it does offer the excitement of evaluating new merchandise, attending fashion shows, and traveling to foreign markets, it also requires considerable time commitment and technical skills. In today's retail environment buyers are also involved in product development. Although a sense of style and color is a necessity, the ability to make quantitative decisions is the utmost important qualification.

Fashion Director (时装总监) This high-level position in most major department stores often carries the title of vice president. The fashion director studies the fashion industry day to day, so that the store is prepared to accommodate any fashion innovation. [4] Many stores use fashion shows to promote their merchandise. Often it is the fashion director who plans the show's format, pulls the merchandise from the different departments, hires models and musicians, prepares the program, and arranges the seating plans.

Advertising Manager (广告经理) The advertising managers must shape the image of the store's advertising campaigns. Usually they are graduates with majors in advertising or graphic arts.

Visual Merchandiser (视觉营销设计师,陈列师) Individuals who plan window and interior displays, determine the best way in which merchandise should be presented on the selling floors, and have significant input into store design are called visual merchandisers. [5]

Market Consultants and Fashion Communications
市场咨询和时尚传播领域

Retailers, manufacturers, and designers are always interested in having as much information as possible so that both short- and long-term goals are satisfactorily achieved. Throughout the fashion industry consulting companies and some trade papers and magazines function specially to help such clients. The following two positions are the important ones in these two fields.

Fashion Forecaster(流行预测员) Fashion forecasters visit the textile mills to assess the fabrics and colors that will be featured in clothing approximately 12 to 18 months later, study haute couture designs that will probably be translated into more affordably priced models, analyze social, political, and economical events that could become the basis of future fashion trends. [6] Often, they travel to foreign countries to observe the way in which people there dress—this is sometimes the inspiration for new designs. *A fashion forecaster's career requires good verbal and writing skills, a keen, understanding of fashion fundamentals, and the ability to participate in research endeavors.* The salaries for such individuals are high, because their forecasts often become the basis for future business decision.

Fashion Editor(时尚杂志编辑) To be the fashion editor of such magazines as Vogue, GQ, Harper's Bazaar, Ebony, YM, or Glamour is to have reached the pinnacle of success in fashion communication. These positions not only offer significant monetary rewards, but also provide an element of excitement. Many editors are extremely influential in helping to promote or destroy a fashion concept. Designers and manufacturers generally try to befriend these people in the hope that it will bring them good press. The requirements of such a position include the ability to write.

Highlights

- The fashion industry offers a variety of career opportunities in textiles, manufacturing, designing, retailing, marketing and communications.

- The fashion industry accommodates people with diverse educational backgrounds. There are, however, some technically oriented careers that require a mastery of specific skills.
- Throughout the fashion industry, consulting companies and some trade papers and magazines function specially to help retailers, manufacturers, and designers acquire as much useful information as possible.

For Review

1. In what situation would a manufacturer employ a stylist?
2. Why is it necessary for patternmakers to understand the use of a digitizer?
3. Describe the role of fashion director for a major retailer.

New Words and Expressions

rigorous *adj.* 严格的,缜密的
credential *n.* 资格证书,资格
format *n.* 设计格式
repeat *n.* (花纹)循环,完全组织
remuneration *n.* 报酬,补偿
colorist *n.* 配色师,色彩设计师,着色师
utilization *n.* 利用,应用
grapher *n.* 制图员
graph *n.* 制图,图表
converter *n.* 布料加工批发商,制衣加工商
gray goods *n.* 坯布,本色布
dyer *n.* 染色师
stock *n.* 原料,配料

care *n.* 保养,维护
drape *n./v.* 悬垂性
draping *n.* 立体剪裁
main stay *n.* 支柱
merchandiser *n.* 经营商,贸易商
stylist *n.* 风格师,造型师,服装搭配师
grader *n.* 推板师
digitizer *n.* 数字转换器,数字化设备
cutter *n.* 裁剪师;裁剪机
plague *v.* 困扰,烦恼
markup *n.* 提价幅度;标价
accommodate *v.* 提供食宿;迎合,迁就
consultant *n.* 顾问;咨询员

Translation

Translate the following sentences italicized in the text into Chinese.

1. Although formal education and training are beneficial, successful people in fashion may have studied fine arts, marketing, design, textiles, or just a broad-based program.

2. The textile industry employs technically skilled individuals, whose talents and

abilities range from artistic creativity to production and sales.

3. Dyeing, printing, and the application of a variety of fabric finishes constitute converting.

4. Designers are responsible for setting the tone of a line in terms of silhouette, color, fabrication and trim.

5. They allocate dollars to be spent on each division's merchandise offerings and instruct the divisional merchandise managers about the store's markup and profitability goals, image, and other policies with a merchandising orientation.

6. A fashion forecaster's career requires good verbal and writing skills, a keen. understanding of fashion fundamentals, and the ability to participate in research endeavors.

Websites

By accessing these Websites, you will be able to gain broader knowledge and up-to-date information on materials related to this chapter.

Fashion Careers:

http://www.fashion-careers.com

Monster. com:

http://www.monster.com

Notes to Part A

[1] Textile designers are artists who create particular patterns and present them in a format that can be translated into fabrics. They paint their designs on paper or fabric and prepare the repeats that will be used in the finished products.

织物设计师是艺术家,他们创造有特色的图案,并将其以某种形式转移到织物中。他们把设计图案绘制在纸上或织物上,并准备将这些图案印制在最后的成品中。

[2] At the very core of the fashion industry is the manufacturer. Whatever the products, manufacturing positions cut across all lines.

服装生产商是整个时装业的中心环节。无论生产何种产品,服装生产中的岗位都要涉及所有生产环节。

[3] A thorough knowledge of textiles and color is necessary, so that the stylist can substitute fabrics and colors in the original designs to make the copy cost-

efficient.

风格师必须精通纺织品和色彩,只有这样,他们才能够选择合适的面料和色彩来替换原创设计的面料及色彩,以降低成本。

[4] The fashion director studies the fashion industry day to day, so that the store is prepared to accommodate any fashion innovation.

时装总监时刻关注服装业的变化,以便商店随时准备好适应任何流行的新鲜事物。

[5] Individuals who plan window and interior displays, determine the best way in which merchandise should be presented on the selling floors, and have significant input into store design are called visual merchandisers.

视觉营销设计师(陈列师)安排橱窗和店内商品的展示,确定在出售地点陈列商品最好的方式。

[6] Fashion forecasters visit the textile mills to assess the fabrics and colors that will be featured in clothing approximately 12 to 18 months later, study haute couture designs that will probably be translated into more affordably priced models, analyze social, political, and economical events that could become the basis of future fashion trends.

时尚预测工作者,通过参观纺织厂来预测未来一年甚至一年半以后的织物及色彩特点;通过研究各种高级时装设计来预测消费者能买得起的时装样式,并且分析可能会影响未来时尚流行趋势的社会、政治及经济事件。

Supplementary Reading
Case Study 1

A Successful Job Candidate

Stacey Peters will graduate from college in 2 months with a degree in fashion merchandising. In addition to the required liberal arts courses, she has studied a number of fashion-related subjects, such as retail management, textiles, advertising, fashion coordination, and fashion publicity. Her cumulative grade point average is 3.7 out of a possible 4.0, which will enable her to graduate with honors. Because she comes from a family that has been in the fashion industry at various levels, Stacey brings a great deal of enthusiasm(热情) to a prospective employer.

Like any diligent prospective graduate, Stacey has done her investigative homework. She has researched numerous manufacturing companies and retail organizations, preparing a resume appropriate to each. To her credit, her time and effort have paid off. She was invited to interview with three prestigious fashion-oriented retail organizations and five manufacturers. The interviews resulted in three firm offers:

(1) Smith and Campbell, a department store with 15 branches, offered her admission to their executive training program, which could lead to either a merchandising or management career. The starting salary is $25000 with future raises based on her ability to perform.

(2) Design Images, a high-fashion chain organization, offered her a position as an assistant store manager in one of the company's 35 units. For a starting salary of $21000, Stacey would assist the store manager and have such decision-making responsibilities as employee scheduling, visual merchandising, handling customer complaints, and inventory replenishment(存货补给). Within 2 years she could become a store manager.

(3) The Male Image, a men's wear designer and manufacturer, has agreed to hire Stacey as a sales representative. Initially she would sell the company's line in the showroom for a salary of $27000 and eventually become a "road" salesperson with compensation based on straight commission(佣金).

Each of the companies is based in Stacey's general geographical area and provides potential for a successful career.

Questions

1. What aspects of each job should Stacey investigate before making a decision?
2. What are the drawbacks that might arise at each job?
3. Which position would you suggest that she accept? Why?

Case study 2

Diana Vreeland

It was the outbreak of World War I that brought 8-year-old Diana Vreeland from Paris to the United States. In Paris, she had lived in a world in which art, culture, and fashion played dominant roles. Her parents knew such people as Diaghilev and Nijinsky. As a result Vreeland felt comfortable mingling in a society that was open to only a few.

In 1924 she married Thomas Vreeland. When they moved to New York in

1937, she accepted a position at Harper's Bazaar that was offered to her by then fashion editor, Carmel Snow.

Although most people outside of the industry believe that a sure ticket to the inner circles begins with "natural beauty", Vreeland often regarded herself as an ugly duckling. Lacking the conventional beauty often associated with those in fashion, she chose to create a persona(人格面具) that epitomized(表现) individuality and style. She had short black hair, rouged cheeks, and bright red lips that set her apart from the rest. Her writing style was as unique as her dress. In 1939 she became fashion editor, working with Mrs. Snow and art director Alexey Brodovitch. In 1963, she left Bazaar to work as associate editor at Vogue; she became editor-in-chief, a position she held until 1971.

Unlike those who merely reported on the fashion scene, Vreeland was a significant promoter. Whenever she felt something was important, she prominently placed it in her column.

From 1972 to 1989 she served as consultant to the Costume Institute of the Metropolitan Museum of Art. Exhibits such as "Balenciaga", "Yves Saint Laurent", and "The Glory of Russian Costume", to name a few, were among the fashion subjects covered during her tenure at the museum. She was considered to be one of the leading players ever to work in the fashion industry.

Questions

1. Why did Diana Vreeland regard herself as an ugly duckling?
2. How did Diana Vreeland first enter the fashion field?
3. What factors contributed to Diana Vreeland's success as a fashion editor?

Part B

Recruitment Advertisement
招聘广告

英文招聘广告因划分标准不同,产生的类别也随之不同。如按广告应征职位的多少,可将招聘广告分为:多部门、多职位招聘广告,单部门、多职位招聘广告,单项职位招聘广告;按广告所占版面的大小,可把招聘广告分为:长篇广告、中篇广告和短篇广告等。不过无论哪种类别的招聘广告,都大致包括以下几项内容:招聘单位简介,对应聘者的资格要求(包括应聘者的年龄、工作经验、学历、所聘工作的职责以及能够享受的待遇等),联系方式(如邮寄或寄电子邮件等方式)。英文招聘广告,语言简明扼要,经常使用缩略词。

Sample 1

下面是一家英国跨国服装企业的一则单项职位招聘广告的范例。

HEAD OF PURCHASING (Europe)

Fashion Industry International

C. ￡25K package.[1]

As one of the United Kingdom's Fashions, and part of a major UK Group, this ￡45m turnover company is seeking to appoint a person of exceptional caliber as Head of Purchasing (Europe).[2]

The principal function of the appointee will be to take overall responsibility for the purchase of all type ladies' fashion goods from European manufacturers. The Head of Purchasing will work within a substantial budget and will be responsible for optimizing purchase prices.[3]

Ideally the successful candidate will be 30~45 years of age, of good educational standard and competent in English and at least two European languages.

Successful experience in trading in the fashion market, of controlling a purchase budget of ￡20m plus, and of weighing risks against potential profit is required.[4] Only candidates with proven negotiating skills, numeracy, self-motivation, and excellent personal and communication skills need apply.[5]

The salary package includes a generous travel and expense allowance, pension scheme, free sickness insurance and promotion opportunities.[6]

Mr. J. Gullen, Price Bowers plc, St.

Michael's Court, Green Street,

Stockport M62 2JF. Ref: M37410/EG.

New Words and Expressions

turnover 营业额	proven negotiating skills 具有一定的谈判技巧
exceptional 杰出的,优秀的	
caliber 能力,才干	numeracy 计算能力
appointee 受聘者	package 一揽子交易,一整套东西
optimize 使最优化	allowance 津贴,补助
weigh 权衡,盘算,仔细考虑	pension scheme 养老金计划

Notes

［1］C. ￡25K package.

薪资约为 25000 英镑,其他福利另加。C. (= about)；K 是 1000 的缩略词。

［2］This ￡45m turnover company is seeking to appoint a person of exceptional caliber as Head of Purchasing(Europe).

这个年营业额达 4500 万英镑的公司,现正物色一名有特殊才能的人士担任公司的欧洲采购主管。

［3］The Head of Purchasing will work within a substantial budget and will be responsible for optimizing purchase prices.

采购主管要进行大额的预算,并且负责拿出最优的采购价。

［4］Successful experience in trading in the fashion market, of controlling a purchase budget of ￡20m plus, and of weighing risks against potential profit is required.

应聘者需要有服装市场贸易、管理 2000 万英镑以上的采购预算以及对潜在利润构成的风险进行估算的成功经验。

［5］Only candidates with proven negotiating skills, numeracy, self-motivation, and excellent personal and communication skills need apply.

应征者还应具备一定的谈判技巧、计算能力以及自我激励、良好的人际交往和沟通能力。

［6］The salary package includes a generous travel and expense allowance, pension scheme, free sickness insurance and promotion opportunities.

待遇包括优厚的旅行费补贴、退休金计划、免费的疾病保险和晋升机会。

Sample 2

PERSONAL ASSISTANT / SECRETARY

Required for Chief Executive of a new company involved in film and television industry.

Requirements Include:

- Ability and initiative to work on your own and to develop the position to its full potential.
- Good administrative and secretarial skills.

- Experience in the film and television industry would be useful, but not essential.
- Good salary, negotiable according to age and experience.

Please reply with typed C. V. [1] to:

Mr. Cynthia Dryden,

44 Randolph Square,

London ECI 4BE

Notes

[1] C. V. (curriculum vitae) 简历,这是英国英语。还可用 résumé。

Sample 3

Senior 3D Technical Designer

DESCRIPTION

Our team is looking for an entrepreneurial, analytical, and highly motivated Senior 3D Technical Designer with prior Technical Design or Patternmaking experience. This role will ensure our products delight our customers in terms of product quality and fit. They will be part of a team of experts who will work with designers and merchandisers to ensure we build best in class products.

As a Senior 3D Technical Designer you will support the design process by creating 3D apparel and photorealistic images. These digital apparel prototypes will be used for concept review, material and color specifications and transferred to partner factories for sample creation. 3D prototypes must convey all necessary aesthetic and technical details while following milestone calendar objectives. The 3D Designer will collaborate with apparel design and brand management teams and participate in the product creation process.

The ideal candidate will be a self-starter with a passion for delivering high quality products to customers. This candidate will possess partner management skills, project management skills, a high degree of ownership and integrity, a high attention to detail, excellent communication skills, and be a great team player. The successful candidate must have a demonstrated ability to render apparel product in 3D and have a solid understanding of patternmaking and garment construction. Candidates should have prior experience in 3D rendering, fit, pattern making, size grading, and garment construction with a top tier branded or vertical fashion retailer.

Primary Responsibilities:
- Leverage 3D software to create and revise 3D apparel renderings.
- Build 3D block library from existing 2D pattern library to support product development process.
- Develop and maintain a virtual fit protocol and standard.
- Develop, modify, and maintain 3D Avatars to support all soft lines apparel categories.
- Manage and maintain a virtual fabric library.
- Liaise with Design and Brand Management teams to provide support for key meetings and milestones.
- Research and apply emerging 3D technology to the product development process.
- Drive for continual improvement and innovation; sponsor newness.
- Evolve Technical Design strategy, operations, methods, and measures of effectiveness.
- Research and resolve fit, pattern, and grading issues in partnership with the Technical Design team.
- Build audit mechanisms to consistently monitor and check fit, grading, and vendor compliance.
- Collaborate with other members of the team on both ongoing and new products.
- Support the Technical Design team in meeting seasonal deliverables and projects.
- Travel to sourcing offices and/or factories to troubleshoot, educate, and build factory capabilities.

Basic Qualifications
- Bachelor's Degree required, preferably in Technical Design, Fashion Design, or Patternmaking.
- 2-3 years of 3D Apparel rendering experience.
- 5+years of Apparel Experience preferably with knowledge of technical design and pattern making.

Preferred Qualifications

- Exceptional interpersonal, organizational and communication skills.
- Ability to work in an extremely creative, fast paced team environment.
- Proven track record for cultivating strong relationships with key client and supplier stakeholders that have resulted in high customer satisfaction.
- Strong analytical skills including the ability to distill, synthesize, and draw conclusions on large amounts of data.
- Ability to manage multiple simultaneous projects requiring frequent communication, time management and problem solving skills.

New Words and Expressions

entrepreneurial 企业家的,创业者的;中间商的
photorealistic 真实感;逼真;拟真
prototype 原型;雏形
render 渲染
avatar 虚拟化身,虚拟模特
liaise 联络,联系
troubleshoot 故障排除

Keys to Translation

Chapter 1

1. The world of fashion began with individual couturiers and evolved, as a result of the Industrial Revolution, into a mass-market industry.

时装业最初起源于个体高级时装设计师,后来由于工业革命,发展成现在的大众市场的产业。

2. During the early 19th century, the opulent designs that dominated the wardrobes of the rich began to disappear; less elaborate dress became the order of the day.

到了19世纪早期,曾经主导富人衣柜的豪华设计逐渐消失;而相对简单的女装开始大行其道。

3. During this time, the Western world witnessed the growth of the middle class, which prospered from new avenues of trade and industry, and spent money on such luxuries as fine clothing.

这一时期,中产阶级在西方世界兴起,他们通过做生意和办企业发家致富,并且愿意花钱买高级服装等奢侈品。

4. Because of the competitive advantage these inventions gave to manufacturers, England was very protective of its discoveries, and forbade the emigration of textile workers and the exportation of its textile machines.

由于这些发明使得英国的生产商具有很大的竞争优势,因此英国政府积极保护这些发明,禁止纺织工人移民他国和纺织机械的出口。

5. With this invention, women began to sew professional-looking clothes at home, and factories experienced the birth of ready-to-wear apparel.

缝纫机的发明使得妇女在家就可以缝制出像专业人员制作的服装,现代成衣业也由此诞生了。

6. Although globally renowned designers are still crowding the runways with outrageous styles at prices that only a few can afford, new designers are showing fashions that reflect what is taking place on the streets, in the political arenas, in the entertainment field, and in movements to protect the environment.

尽管世界顶级的设计师仍然使得 T 台上充斥着只有少数人可以承受的、价格不菲的高级时装样式,但是新一代的设计师却开始展示反映街头时尚、反映政治界、娱乐界的风云变幻,甚至是环保运动的时装。

Chapter 2

1. Although we are taught that we should not judge others by their appearance alone, relying on appearance to guide personal decisions and social interactions is not only natural, but also inescapable.

尽管我们所受的教育是不应单纯地以貌取人,但是凭借外表来对他人进行判断和进行社交不仅是很自然的,而且是不可避免的。

2. Most people would like to improve their appearance with appropriate clothing by camouflaging their less desirable attributes and highlighting the more attractive aspects of their bodies.

多数人喜欢通过选择合适的着装来改善自己的外观形象,掩饰自己身体不尽如人意的地方,并且彰显自身的优势。

3. One school of thought is that "beauty is in the eye of the beholder", that individual attraction is a result of personal experience, cultural background and specific circumstances.

一种思想流派认为"情人眼里出西施",也就是说,所谓个人的吸引力是一个人的经历、文化背景以及特殊的环境因素综合作用的结果。

4. Ancient Greeks believed that the world is beautiful because there is a certain measure, proportion, order and harmony between its elements.

古代希腊人认为,世界之所以是美的,是由于组成它的各个部分之间存在着某种尺寸标准、比例、秩序与和谐。

5. There is also a greater difference in the depth ratios from front to back in the female figure than there is in the male with respect to bust/waist and waist/hip relationships.

在胸围/腰围比例,以及腰围/臀围比例方面,男性与女性的身体存在着很大的不同。女性身体的厚度要比男性大。

6. The appearance of the clothed body is a perception of the viewer (whether of the wearers themselves or others) in a social and climatic context.

着装的身体外观是处于一定社会和风气中观察者的知觉表现。这一点无论对于穿着者本身还是其他人而言都是一样的。

Chapter 3

1. Information on color trends for the coming season usually comes from fiber companies and professional color services like Pat Tunskey.

有关下一季的色彩流行趋势的信息,通常来自纤维公司和像 Pat Tunskey 这样的专业色彩服务公司。

2. The refabrications and versions of good bodies from the last season may provide the basic styles or staples of the new line.

上一季销量较好的款式改用别的布料可能成为新产品的基本款式或主要服装。

3. Most customers touch a garment immediately after being visually attracted to it, so the hand (feel) of the fabric is as important as its appearance.

多数消费者被一件衣服吸引之后都会立刻用手去触摸它,因此,服装面料的手感和其外观一样重要。

4. When the designer styles a garment, he or she must consider the price of every detail from the initial cost of the fabric to trims and construction methods.

设计师设计服装时必须考虑每个细节的价格,从最初的布料成本到配饰成本和服装生产成本。

5. High-priced garments usually require dry-cleaning because they are made of delicate fabrics and there is little customer resistance to the maintenance cost.

高价位的服装由精致的织物做成,因此通常需要干洗,而且消费者也愿意在衣服保养方面花钱。

6. The silhouette or shape of the garment is important because it tells the patternmaker how much fullness should be in a skirt, sleeve or bodice.

服装的轮廓线或外形很重要,因为它能使样板师知道在裙子、袖子或者上衣片等地方的宽松度是多少。

Chapter 4

1. Some designers review lines with either a stylist or the owner of the house, but most manufacturers encourage the designer to make aesthetic decisions about fabric.

某些服装设计师会同服装搭配师或老板一起选择织物,但更多的服装生产厂商鼓励设计师根据自己的审美观来选择织物。

2. Generally, piece-goods salespeople encourage manufacturers who are buying substantial amounts of yardage to commit (buy a specific amount or a particular fabric) early in the season to ensure delivery at the promised time.

通常，布匹销售商鼓励购买数量大的生产厂家在季节刚开始时提出所需要的数量或特定织物，以确保按时交货。

3. A designer needs to know who else has sampled a fabric and will try to avoid fabrics that have been chosen by competitors or those used in lower-priced lines.

设计师需要了解别人已选择的布样，以尽可能地避免采用竞争对手已选择的布料或低档廉价服装使用的布料。

4. A designer should see as many lines as possible. The piece-goods salespeople are well acquainted with events in the marketplace, and they often have information that can aid a designer in the choice of a fabric or print.

设计师应该尽可能多地去看产品。布匹销售商非常熟悉本行业的市场，他们通常拥有能帮助设计师选择布料和花色的信息。

5. The cardinal rule for a designer should be this: see all the textile representatives you possibly can during a season; judge their products' relevance to the designs you are planning; and be flexible, yet practical.

设计师需要遵守的一条重要规则是：在一个季节里尽可能多地接触纺织品代理商，判断他们的产品是否可以用到你未来的设计中。要灵活，但更要实际。

6. Quick delivery of an acceptable replacement is critical, and the designers often shop the fabric market looking for piece goods for the line in production as well as sampling new fabrics for the coming season.

快速运送可接受的替代面料非常重要。设计师经常光顾纺织品市场，既为正在生产的服装系列寻找面料，也为下一季的设计样品寻找新的面料。

Chapter 5

1. Measurements in square brackets after each step are for the basic 91.4 cm (36 in) block.

每项后面方括号中的数据，为91.4 cm(36 in)的原型板数据。

2. Approximately 25.4 cm (10 in) down from top of paper and 20.3 cm (8 in) in from left, draw a horizontal line AB, half he hip measurement [49.5 cm(19.5 in)].

距纸上边约25.4 cm，距左边约20.3 cm，画一条水平线AB，其长度为臀围的一半[49.5 cm (19.5 in)]。

3. Square up to waist level 20.3 cm(8 in). Draw across C to D.

(从 A、B 点)垂直上量 20.3 cm(8 in)得 C、D 点,过 C、D 点画直线。

4. The dart size depends on the difference between amount in waist now and amount needed.

省的大小依赖于腰围量和实际需要量的差量。

5. Trace the trouser pattern to the point where it touches the CB or CF skirt lines. Use skirt darting and waistline.

绘出裤子板型到一点,这点交到裙子的 CB 或者 CF 线上。使用裙子的省和腰线。

6. Re-position traced slack piece to the other side of pleat. Draw around complete culottes pattern.

重新将绘制的宽松裤的裤片摆放到褶的另一边。完成整个裙裤样板。

Chapter 6

1. A businessperson with a good idea or the ability to sell a product can capitalize on his or her talent and pay small-business people to complete the manufacturing process.

具有良好创意或能力的企业家能够充分利用他或她的智慧,雇佣小公司的人去完成生产过程。

2. It markets the line produced by the design department and acts as an intermediary between buyers and the designer.

它(销售部)销售设计部门所生产的产品,同时作为买家和设计师之间的中间人。

3. The manufacturer must carefully weigh the advantages of inexpensive labor against shipping costs, customs fees and the cost of coordinating the offshore production.

生产商必须认真权衡廉价劳动力相对于运输费、关税和国外生产协调的费用之间的利弊关系。

4. Increased flexibility resulting from the ability of these manufacturers to cut smaller lots and deliver rapidly allow stores to carry smaller inventories and makes an American-made garment more desirable.

这些生产商能小批量裁剪,交货迅速,从而增强了其灵活性,使商店可以保持较小的库存,使美国制造的服装更受欢迎。

5. When they find the savings are substantial, interlocking systems that will further speed production and reduce costs are added to the basic computer.

当发现大幅节约了成本时,他们就在其基础计算机系统上添加了可以进一步提高生产效率、降低成本的连锁系统。

6. System producers estimate that a minimum savings of 3 percent on fabric alone is typical for computerized markers, a significant factor since fabric is the most expensive element of a garment's cost.

系统的生产商估计,仅用计算机排料就至少可以节约3%的布料,这是很可观的,因为布料往往是服装成本中最贵的部分。

Chapter 7

1. Fashion marketing is the entire process of research, planning, promoting and distributing the raw materials, apparel and accessories that consumers want to buy.

时装市场营销是包括对消费者需要的原料、服装和配饰所进行的全部市场调研、企划、促销以及销售活动的全过程。

2. Credit, productivity, inflation, recession, labor costs, and international currency values affect purchasing power.

信贷、生产能力、通货膨胀、经济衰退、劳动力成本和国际货币价值都会影响消费者的购买力。

3. Therefore, ever-increasing amounts of textiles, apparel, and accessories are being imported into the United States and the European Union (EU) because of the availability of cheaper labor in low-wage countries. United States and European manufacturers compete with labor rates of 69 cents per hour in parts of China and 60 cents or less in India.

因此,美国和欧盟对纺织品、服装和配饰的进口都一直在增加,原因在于低工资国家的劳动力成本低。美国和欧洲的生产商们在劳动力价值方面进行竞争,在中国的一些地方劳动力每小时为69美分,而印度则为每小时60美分或更低。

4. Ideally, the two figures should be about equal. Lately, however, the United States has been importing much more than it exports sending American dollars abroad to pay for these goods and creating a huge trade deficit.

理想的情况是进出口的价值持平。但最近美国进口的产品远远大于出口,大量美元外流支付进口产品,出现巨大贸易逆差。

5. Additional functions of the WTO are to set rules governing trade behavior, set environmental and labor standards, protect intellectual property, resolve disputes between members, and serve as a forum for trade negotiations.

另外,世界贸易组织的其他功能包括制定贸易规范,确立环境和劳动力标准,保护知识产权,解决成员国之间的贸易争端,并为商业谈判提供服务。

6. They source (find the best production available at the cheapest price) contract production to suppliers who can handle everything from buying piece goods to booking space on ships.

他们将产品生产外包给能生产出最为价廉物美,并能处理从材料购买到定船运输等所有事宜的供应商。

Chapter 8

1. Retail organizations in the 21st century bear little, if any, resemblance to the trading posts and general stores of earlier days.

21世纪的零售组织与早期的交易站、综合商店即使有相似之处,也非常少。

2. As merchandise became more varied, some early retailers moved away from the general merchandise concept and pioneered the first limited line stores, which concentrate on one merchandise classification.

由于商品越来越多样化,一些早期的零售商摆脱了综合商品销售的概念,首创了局限于某种服装的专卖店,即只销售一个种类的商品。

3. Flea markets have many vendors, each with comparatively limited retail space and lower operating costs than traditional retailers. Vendors work on very low markups and are able to sell at considerably reduced prices.

跳蚤市场有很多卖主。和传统的零售商相比,他们的经营空间有限,营业成本较低。他们的毛利很低,货物能够大幅度降价销售。

4. Not only are most merchandising and policy decisions made in these locations, but also the parent stores generally account for a significant part of the company's sales volume.

不仅大部分的销售策划和决策都在这些市区总部进行,而且母公司(商场)的营业额通常也占公司总营业额的一大部分。

5. The store wraps the gifts, like perfume, small leather goods, and home products, and sends them to designated clients with very little customer involvement.

商场可以包装礼品,比如香水、小型皮革制品和家用产品等,然后把它们送

给指定的客户，几乎不用顾客费心。

6. A major responsibility of the fashion-oriented retailer is to develop an image that will motivate shoppers to become customers. Those who successfully do this become the major players in the game of fashion.

销售定位为时尚商品的零售商，一个主要的责任就是要提升自己的形象，从而促使购物者成为自己的顾客。在这方面做得成功的零售商就成了时尚界的弄潮儿。

Chapter 9

1. Designers, manufacturers, and retailers of fashion merchandise pay as much attention to promoting their products as they do to designing and merchandising them.

服装设计师、生产商和时尚商品零售商既重视对产品促销，也重视设计和销售。

2. One is to make the trade aware of their creations—that is, to capture the attention of the store buyers and merchandisers who make selections for their particular clienteles.

一种目的是让业界了解他们的作品，也就是要吸引那些为特定客户选购产品的商场买手和采购人员的注意。

3. The major share of the retailer's promotional budget is earmarked for advertising; the major portion of the advertising budget is spent on newspaper ads.

零售商的大部分推广预算都要用在广告上；而广告的大部分预算则用于在报纸上刊登广告。

4. Whether the ads are trade oriented and directed toward the industry or consumer based and focused on the ultimate consumer, the formats are either promotional or institutional.

不管广告是定位于贸易，直接面向行业或是以消费者为基础，还是针对最终的消费者，其所采取的形式要么是产品促销性的，要么是公司形象宣传性的。

5. Sometimes fragrance and cosmetic vendors reward purchasers of their products with premiums that are free or comparatively inexpensive.

有时候，香水和化妆品的商家会通过赠送或适当降价卖出其产品来奖励购买者。

6. The presentation can also be produced by consultants, who are paid by

companies to design settings and execute display installations, and by freelancers, who periodically install visual presentations.

视觉展示也可以由公司出钱让顾问来组织、设计产品的陈列和具体实施产品展示。同时,自由职业者也可定期来布置产品的视觉展示。

Chapter 10

1. Although formal education and training are beneficial, successful people in fashion may have studied fine arts, marketing, design, textiles, or just a broad-based program.

尽管正规的教育和职业训练是有益的,但时尚界成功人士可能学过美术、营销、设计、纺织或更宽泛的专业。

2. The textile industry employs technically skilled individual, whose talents and abilities range from artistic creativity to production and sales.

纺织业雇用训练有素的各种技术人才,他们的智慧和能力体现在创造性和艺术性以及生产和销售方面。

3. Dyeing, printing, and the application of a variety of fabric finishes constitute converting.

布匹加工主要包括染色、印花以及织物的各种整理技术的应用。

4. Designers are responsible for setting the tone of a line in terms of silhouette, color, fabrication and trim.

服装设计师负责确定廓型、色彩、织物选用和装饰方面的服装风格。

5. They allocate dollars to be spent on each division's merchandise offerings and instruct the divisional merchandise managers about the store's markup and profitability goals, image, and other policies with a merchandising orientation.

他们(销售总监)负责将促销资金分发给各个分部,并以营销店的商品定价、盈利目标、店面形象和其他有关销售定位的政策对个分部经理进行指导。

6. A fashion forecaster's career requires good verbal and writing skills, a keen, understanding of fashion fundamentals, and the ability to participate in research endeavors.

时尚预测员这一职业要求从业人员具有很好的口头和书面语言表达能力,对影响时尚变化的要素有敏锐的理解能力和参与研究的能力。

Glossary（词汇表）

20(covered) 20个包扣
abandonment[əˈbændənmənt] n. 放弃；拒绝
accommodate[əˈkɔmədeit] v. 提供食宿；迎合，迁就
accounting[əˈkauntiŋ] n. 会计学
adjacent[əˈdʒeisnt] adj. 相邻的，邻近的
adopt[əˈdɔpt] v. 采纳
advertise[ˈædvətaiz] v. 广告，宣传
advising bank 通知行
aesthetic[iːsˈθetik] n. 美学
affluent[ˈæfluənt] adj. 富裕的，丰富的
alliances[əˈlaiəns] n. 联合
allowance[əˈlauəns] n. 津贴，补助
animated[ˈænimeitid] adj. 动画的，栩栩如生的
applicant[ˈæplikənt] n. 申请人
appointee[ˌəpɔinˈtiː] n. 受聘者
appraise[əˈpreiz] vt. 评定，鉴定
approximately[əˈprɔksimətli] adv. 近似地，大约
arbitrary[ˈɑːbitrəri] adj. 随意的，任意的，武断的
arena[əˈriːnə] n. 活动场所
aristocracy[ˌæriˈstɔkrəsi] n. 贵族，权贵阶级
ART(article) 货品，作品
artificial[ˌɑːtiˈfiʃl] adj. 人造的，人工的
assign[əˈsain] v. 分配；分派；指定
assortment[əˈsɔːtmənt] n. 种类，类别
assume[əˈsjuːm] v. 承担
attire[əˈtaiə] n. (文学用语)服装
attribute[əˈtribjuːt] n. 特征，属性
authentic[ɔːˈθentik] adj. 真迹，原作的，真实可靠的
avatar[ˈævətɑː] n. 虚拟化身，虚拟模特
avenue[ˈævənjuː] n. 方法，途径；林荫大道
awl[ɔːl] n. 锥子
base goods n. 基本商品，基本原料

Glossary（词汇表）

be akin to   类似

belt loop   饰带,带襻

beneficiary [ˌbeniˈfiʃəri]   *n.* 受益人

binder [ˈbaində]   *n.* 活页夹,活页本

binding [ˈbaindiŋ]   *n.* 镶边,绲边

bisecting [baiˈsektiŋ]   *adj.* 对角的

block [blɔk]   *n.* 原型,剪裁样板

bodice [ˈbɔdis]   *n.* 女装紧身上衣;上衣片,大身

body dimensions   *n.* 身体尺度,尺寸

bombard [bɔmˈbɑːd]   *vt.* 炮轰

bonus [ˈbəunəs]   *n.* 红利

boulevard [ˈbuːləvɑːd]   *n.* [美]林荫大道

brisk [brisk]   *adj.* 活泼的,活跃的

brisk demand   需求活跃

budget [ˈbʌdʒit]   *n.* 预算 *v.* 做预算

C = commission   佣金

C No.（Carton Number）纸箱号

cable address   电报挂号

caliber   *n.* 能力,才干

camouflage   *vt./n.* 伪装,遮掩,掩饰

capitalize on   利用

cardboard carton   纸箱

cardinal [ˈkɑːdinl]   *adj.* 重要的

care [kɛə]   *n.* 保养

care label   护理标签

category [ˈkætəgəri]   *n.* 类别,种类

cathexis [kəˈθeksis]   *n.* 满意度

celebrity [səˈlebrəti]   *n.* 名人,名声

charismatic [ˌkærizˈmætik]   *adj.* 神赐能力的;超凡魅力的

charitable [ˈtʃærətəbl]   *adj.* 仁慈的,(为)慈善事业的

choreography [ˌkɔriˈɔgrəfi]   *n.* （芭蕾舞或诗歌等）的舞步设计与编排艺术

CIF = Cost Insurance and Freight   到岸价格

CIFC 5%：CIF = cost, insurance and freight   成本、保险加运费

CIFC 5% New York per piece in RMB   纽约抵岸价,包括5%的佣金,以人民币(元)计算,每件

circumference ratios   *n.* 圆周长,这里指身体围度

clientele [ˌkliːənˈtel]   *n.* 客户

161

cobbler [ˈkɔblə] n. 鞋匠

colorist [ˈkʌlərist] n. 配色师,色彩设计师,着色师

commitment [kəˈmitmənt] n. 承付款项

competent [ˈkɔmpitənt] adj. 有能力的,胜任的

confidante [ˌkɔnfiˈdənt] n. 知己,密友

confirmed, irrevocable L/C (letter of credit) 保兑的不可撤销的信用证

congested [kənˈdʒestid] adj. 拥挤的

consignee [ˌkɔnsaiˈniː] n. 收货人

consultant [kənˈsʌltənt] n. 顾问;咨询员

converter [kənˈvəːtə] n. 布料加工批发商,制衣加工商

corset [ˈkɔːsit] n. 紧身胸衣

cost-effective adj. 有成本效益的,划算的

cotton gin n. 轧棉机,轧花机

cotton interlock n. 棉毛布

counterbalance [ˌkauntəˈbæləns] vt. 使平衡,抵消

counterpart [ˈkauntəpɑːt] n. 对应物,极相似的人或物;配对物

country of origin 原产国,起运国

couture [kuˈtjuə] n. 高级时装业

couturier [kuˈtjuəriei] n. 高级时装设计师

credential [krəˈdenʃl] n. 资格证书,资格

credit rating n. 信誉度

crepe [kreip] n. 绉纱,绉布

crisp [krisp] n./adj. 挺括(的),挺爽(的)

crotch length 立裆

CTN (carton) 纸箱

CU. M. 立方米

cue [kjuː] n. 暗示,提示

currency [ˈkʌrənsi] n. 货币

custom-tailored adj. 定制的,定做的

cut-mark 裁剪

cutter [ˈkʌtə] n. 裁剪师;裁剪机

dart [dɑːt] n. 省

database [ˈdeitəbeis] n. [计]数据库,资料库

deduct [diˈdʌkt] v. 减去,扣除

defeat [diˈfiːt] v. 使失败,使受挫折

deficiency [diˈfiʃnsi] n. 缺乏,不足

delivery date 发货日期

delivery [di'livəri]　n. 发货,交货

denim ['denim]　n. 粗斜纹棉布,劳动布;牛仔布

depreciation [di͵priːʃi'eiʃn]　n. 贬值;折旧

description of goods　货名

dictate [dik'teit]　n. /v. 命令,支配,摆布

digitizer ['didʒitaizə]　n. 数字转换器,数字化设备

direct [di'rekt]　n. 直运

discount ['diskaunt]　n. 折扣

disguise [dis'gaiz]　n. 伪装

disseminate [di'semineit]　v. 传播

distribution [͵distri'bjuːʃn]　n. 分配;销售

documentary Letter of Credit　跟单信用证

dominant ['dɔminənt]　adj. 有统治权的,占优势的,支配的

down jacket　羽绒夹克

DOZ(dozen)　一打(12件)

draft at sight　即期汇票

drafting table　n. 绘图桌

drape [dreip]　n. /v. 悬垂性

draping ['dreipiŋ]　n. 立体剪裁

drawee [drɔː'iː]　n. 付款人

due [djuː]　adj. 应付的,到期的;约定的

duplicate ['djuːplikeit]　n. /v. 副本;备份

dyer ['daiə]　n. 染色师

earmark ['iəmɑːk]　v. 指定(款项等的)用途

earned labor　工费

ease [iːz]　n. 松量

elaborate [i'læbrət]　adj. 详尽而复杂的,精心制造的

electronic [i͵lek'trɔnik]　adj. 电子的

embellish [im'beliʃ]　v. 美化,装饰

embroider [im'brɔidə]　v. 绣花

emigration [emi'greiʃn]　n. 移民

emporium [em'pɔːriəm]　n. 商场,商业中心;大百货商店

enclosed [in'kləuzd]　adj. 被附上的

endorsed in blank [in'dɔːzd]　空白背书

entice [in'tais]　v. 诱惑,诱使

entrepreneur [͵ɔntrəprə'nəː]　n. 企业家

entrepreneurial [͵entrəprə'nuːriəl]　adj. 富于企业家精神的

estimate[ˈestimət]　v. 估价

evoke[iˈvəuk]　v. 引起，唤起（记忆、感情等）

exceptional[ikˈsepʃənl]　adj. 杰出的，优秀的

exchange rate　n. 汇率

exclusive[ikˈskluːsiv]　adj. 唯一的；高级的

executive[igˈzekjətiv]　n. 执行者，经理主管人员

expiry[ikˈspaiəri]　n. 中止，满期

facility[fəˈsiləti]　n. 设备，工具

fancy[ˈfænsi]　n./adj. 花式（的）；时兴的纺织品（或服装）

fantasy[ˈfæntəsi]　n. 幻想

fashion forecaster　n. 流行预测员

favorable[ˈfeivərəbl]　adj. 赞成的；有利的；起促进作用的

feature[ˈfiːtʃə]　v. 以……为特色

feedback[ˈfiːdbæk]　n./v. 反馈

fiber[ˈfaibə]　n. 纤维

finishing[ˈfiniʃiŋ]　n. 后整理

flagship store　旗舰店

flexibility[ˌfleksəˈbiləti]　n. 灵活性，机动性

fluctuations[ˌflʌktjuˈeiʃəns]　n. 波动

flyer[ˈflaiə]　n. （广告）传单

flying shuttle　n. 飞梭

FOB price　离岸价格

forbid[fəˈbid]　v.（forbade, forbidden）禁止，不许

formula[ˈfɔːmjələ]　n. 公式，规则

franchise[ˈfræntʃaiz]　n. 产品经销特许权

freelancer[ˈfriːlɑːnsə]　n. 自由职业者

format[ˈfɔːmæt]　n. 设计格式

fullness[ˈfulnəs]　n. 丰满度

function[ˈfʌŋkʃn]　n. 函数；功能

fuse[fjuːz]　n. 熔化；融合

fusion[ˈfjuːʒn]　n. 熔化；合成，合并

G. W.（gross weight）　毛重

gabardine[ˌgæbəˈdiːn]　n. 华达呢

garment[ˈgɑːmənt]　n. 服装

Geneva　n. 日内瓦城（瑞士西南部城市）

globalization[ˌgləubəlaiˈzeiʃn]　n. 全球化，全球性

grader[ˈgreidə]　n. 推板师

Glossary（词汇表）

gradient［ˈɡreidiənt］ *n.* 斜度,坡度

grading［ˈɡreidiŋ］ *n.* 放缩板,推板

graph［ɡrɑːf］ *n.* 制图,图表

grapher［ˈɡræfə］ *n.* 制图员

graphic［ˈɡræfik］ *adj.* 绘图的,图表的

gray goods *n.* 坯布,本色布

gross profit calculation 毛利计算

gross［ɡrəus］ 缩写为 gr.,gro. 罗（计数单位,等于144个,12打）

guarantee［ˌɡærənˈtiː］ *v.* 保证,担保

hanger loop 吊襻

hangtag sticker 吊牌贴纸

hard goods （小汽车、电视机等）耐用品

hemline［ˈhemlain］ *n.* （裤边、袖口等）边缝线

highlight［ˈhailait］ *vt.* 使……显著(突出)

hook-up［ˈhukʌp］ *n.* 连接装置,联网,连接

hosiery［ˈhəuziəri］ *n.* 针织品

hourglass［ˈauəɡlɑːs］ *n.* 沙漏

import label 进口标签

in compliance with 遵照

in execution of 执行

in reply 作为回答

in words 大写

index［ˈindeks］ *n.* 指数

inferior quality 质量次的

initial［iˈniʃl］ *n.* 用自己姓名的首字母签字于(某处)

inspection［inˈspekʃn］ *n.* 检查

inspriation［inspəˈreiʃn］ *n.* 灵感

installation［ˌinstəˈleiʃn］ *n.* 安装；装置

institutional［ˌinstiˈtjuːʃənl］ *adj.* 制度上的

interlock［ˌintəˈlɔk］ *adj.* 双罗纹的

interlocking［ˌintəˈlɔkiŋ］ *adj.* 连锁的,联合的

intermediary［ˌintəˈmiːdiəri］ *n.* 调解者；中间者

Intranet［ˈintrənet］ *n.* 企业内部互联网

inventory［ˈinvəntri］ *n.* 详细目录,存货

Invoice No. (Invoice Number).［ˈinvois］ 发票号

irrevocable［iˈrevskəbl］ *adj.* 不可撤销的

issue［ˈiʃuː］ *v.* 开证

jobber [ˈdʒɔbə]　n. 批发商；经纪人

kit [kit]　n. 成套工具，用具包，工具箱，成套用具

kraft bag　牛皮纸袋

L/C NO. (letter of credit)　信用证号

last [lɑːst]　n. 鞋楦

lease [liːs]　v. 租用，租赁

leather [ˈleðə]　n. 羽绒

liaise [liˈeɪz]　v. 联络，联系

license [laisns]　n. 许可证，执照

line [lain]　n./vt. 型，款式，轮廓（线），（商品品种的）系列；给（衣服）装衬里

lining [ˈlainiŋ]　n. 里衬

lodge [lɔdʒ]　v. 提出

log [lɔg]　v. 记录 [login (计算机) 输入指令开始]

long-term interests　长远利益

Lycra [ˈlaikrə]　n. 莱卡

mainstay [ˈmeinstei]　n. 支柱

manufacturer [ˌmænjuˈfæktʃərə]　n. 生产商

mark [mɑːk]　n. 标志

marking [ˈmɑːkiŋ]　n. 划样；排料

mark up [ˈmɑːkʌp]　n. 提价幅度；标价

mass-market [ˌmæsˈmɑːkit]　adj. 大众市场

MEAS (measures)　n. 体积

measure [ˈmeʒə]　n. （估价、判断事物的）尺寸；标准

merchandise [ˈməːtʃəndais]　n./v. 商品，货物；推销，经营

merchandiser [ˈməːtʃəndaisə]　n. 经营商，贸易商

mill No.　厂号

milliner [ˈmilinə]　n. 女帽商

moderately priced　价格适中的

modify [ˈmɔdifai]　vt. 修改，修饰

multimedia [ˌmʌltiˈmiːdiə]　n. 多媒体

myth [miθ]　n. 神话

N.W. (net weight)　净重

navy [ˈneivi]　adj. 海军蓝的，深蓝色的

non-woven [ˌnʌnˈwəuvn]　n. 无纺布

numeracy [ˈnjuːmərəsi]　n. 计算能力

offshore [ˌɔfˈʃɔː]　adj. 海外的，国外的

operation [ˌɔpəˈreiʃn]　n. 加工

Glossary（词汇表）

optimize [ˈɔptimaiz] v. 使最优化

opulent [ˈɔpjulənt] adj. 富裕的,豪华的

orientation [ˌɔːriənˈteiʃn] n. 方向,倾向性;定位;向东方

outrageous [autˈreidʒəs] adj. 不寻常的,骇人听闻的

outstanding [autˈstændiŋ] adj. 未完成的,未付款的

overall [ˈəuvərɔːl] n. 外衣;工装服

overhead [ˌəuvəˈhed] n. 房租、电费等管理费用

overlay [ˌəuvəˈlei] v./n. 覆盖;覆盖图

package [ˈpækidʒ] n. 一揽子交易,一整套东西

packing [ˈpækiŋ] n. 包装

paper patterns n. 纸样

partnership [ˈpɑːtnəʃip] n. 合伙,合股;合伙企业

patent [ˈpætnt] n. 专利

patronize [ˈpætrənaiz] v. 光顾,惠顾

patternmaker [ˈpætən,meikə] n. 样板师

Pc（piece） n. 件

Pcs 件数

pension scheme 养老金计划

per [pə(r)] prep. 经,由

perception [pəˈsepʃn] n. 感知,知觉

periodical [ˌpiəriˈɔdikl] n./a. 期刊,杂志;周期的,定期的

personality [ˌpəːsəˈnæləti] n. 个性,人格;人物,名人

philosophy [fəˈlɔsəfi] n. 哲学,哲学体系;理念

photorealistic n. 真实感;逼真;拟真

physique [fiˈziːk] n.（尤指男性的）体格,体形

piece goods 布匹,匹头

pioneer [ˌpaiəˈniə] v. 创办;打开(新领域)或准备(途径)

piracy [ˈpairəsi] n. 盗版,侵犯版权

PKGS（packages） 包装

placement [ˈpleismənt] n. 安置,布局

plague [pleig] v. 困扰,烦恼

plastic bag sticker 塑料袋贴纸

pleat [pliːt] n. 褶,褶状物

plotter [ˈplɔtə] n. 绘图机

poly bag 合纤袋

polyester [ˌpɔliˈestə] n. 涤纶,聚酯

port of discharge 卸货港

port of loading 装货港

potential [pəˈtenʃl] adj. 潜在的,可能的

power loom n. 动力织机

predecessor [ˈpriːdəsesə] n. 前辈,前任;(被取代的)原有事物

premier [ˈpremiə] adj. 最好的;首要的

premise [ˈpremis] n. [-s]经营场所

prestigious [preˈstidʒəs] adj. 有声望的,受尊敬的

principal [ˈprinsəpl] adj. 主要的,首要的

printed dress 印花连衣裙

prod line 产品线

progressive [prəˈgresiv] adj. 上进的,进步的

prominence [ˈprɔminəns] n. 突出,显著

prop [prɔp] n. 服装等道具

prospective [prəˈspektiv] adj. 预期的,未来的

prototype [ˈprəutətaip] n. 原型;雏形

proven negotiating skills 具有一定的谈判技巧

proximity [prɔkˈsiməti] n. 接近,亲近

putative [ˈpjuːtətiv] adj. 公认的

quota [ˈkwəutə] n. 配额,限额

rayon [reiɔn] n. 粘胶长丝

ready-to-wear n. 成衣

recipient [riˈsipiənt] n. 接受者

refabrication [riˈfæbriˈkeiʃən] n. 重新编造,重复使用,重新构造

reimburse [ˌriːimˈbəːs] v. 偿还,偿付

relegate [ˈreligit] v. 转移;归入;提交

remark [riˈmaːk] n. 备注

remuneration [riˌmjuːnəˈreiʃn] n. 报酬,补偿

render [ˈrendə] v./n. 渲染

repeat [riˈpiːt] n. (花纹)循环,完全组织

reproductive potential 生育潜力

resemblance [riˈzembləns] n. 类同之处

retailer [ˈriːteilə] n. 零售商

rigorous [ˈrigərəs] adj. 严格的,缜密的

run way n. 跑道,时装表演走道

S. S=steam ship 蒸汽轮船

S/O (shipping order) 装货单

sale volume 销售额

Glossary（词汇表）

sample [ˈsɑːmpl] n. 样衣

saturate [ˈsætʃəreit] v. 使饱和,浸透,使充满

scarf [skɑːf] n. 围巾

screen [skrin] v. 筛选

seamless [ˈsiːmləs] adj. 无缝合线的,浑然一体的

security label 安全标签

segment [ˈsegmənt] n. 组成部分

seminar [ˈseminɑː] n. 专题讨论课;短期强化课程

sensory [ˈsensəri] adj. 感官的,感觉上的

shade [ʃeid] n. 色调

shell fabric 面料

shipper [ˈʃipə] n. 托运人

SHIPPING MARKS 装运唛头（进出口货物包装上所做的标记）

side seam 边缝

silhouette [ˌsiluˈet] n. 轮廓

sketch [sketʃ] n. 服装效果图;草图

sloping [ˈsləupiŋ] adj. 倾斜的,有坡度的

soft goods 纺织品,布匹等相关产品

sophisticated [səˈfistikeitid] a. 复杂的

span [spæn] v. 横越,跨越

spandex [ˈspændeks] n. 氨纶

spare button bag 备用纽扣包

specific [spəˈsifik] adj. 具体的,明确的

specification [ˌspesifiˈkeiʃn] n. 规格

spindle [ˈspindl] n. 锭子,纺锤

spinner [ˈspinə] n. 纺纱工,传动齿轮

spinning jenny n. 珍妮纺纱机

sponsor [ˈspɔnsə] n. 发起人,赞助人,主办人 v. 赞助

staple [ˈsteipl] n. 大路货,主要产品

step up the trade 促成交易

stimulus (pl. stimuli) [ˈstimjələs] n. 刺激

stitch [stitʃ] n. 一针,针脚,缝线

stock yardage n. (产品的)原料,库存备料

stock [stɔk] n. 原料,配料

style No. [stail] 款号

style [stail] n/v. 式样,风格

stylish [ˈstailiʃ] adj. 时髦的,漂亮的

169

stylist [ˈstailist]　n. 风格师,造型师,服装搭配师

submit [sʌbˈmit]　v. 提交,递交

substantial [səbˈstænʃl]　adj. 大量的;丰富的;充实的

subtract [səbˈtrækt]　v. 减

supplier [səˈplaiə]　n. 供应商

swatch [swɔtʃ]　n. 样品

tablet [ˈtæblət]　n. 写字板,书写板

tally [ˈtæli]　n./v. 记录;计算

tape [teip]　n. 皮尺

target [ˈtɑːgit]　n. 目标

tariff [ˈtærif]　n. 关税

temper(with) [ˈtempə]　vi. 调和或减轻某事物的作用,缓和

terms of payment　付款方式

term [təːm]　n. 期限;条款

textile [ˈtekstail]　n. 纺织品;纺织原料

texture [ˈtekstʃə]　n. 织物质地;(材料的)纹理,肌理

total direct cost　合计直接成本

total fabric cost　织物总成本

total labor cost　合计人工成本

transshipment [trænsˈʃipmənt]　n. 转运,转船

treadle [ˈtredl]　n. 脚踏板

trial order　试订货

trim [trim]　v. 装饰,镶边

trimming [ˈtrimiŋ]　n. 辅料

troubleshoot [ˈtrʌblʃuːt]　v. 故障排除

turnover [ˈtəːnəuvə]　n. 营业额

turn-time　n. 交货期限

twofold [ˈtuːfəuld]　adj. 两部分的,双重的

under the auspices of　由……赞助,由……资助

undertaking [ˌʌndəˈteikiŋ]　n. 事业,企业;承诺

unit cost　单价

upscale [ˌʌpˈskeil]　adj. (美)高消费阶层的,(商品)质优价高的

utilization [ˌjuːtəlaiˈzeiʃn]　n. 利用,应用

vendor [ˈvendə]　n. 卖东西的小贩;卖主

version [ˈvəːʃn]　n. (同一种物品稍有不同的)样式,复制品,版本

vertical [ˈvəːtikl]　adj. 垂直的,直立的

vertical integration　n. [经]垂直统一管理,纵向整合

Glossary（词汇表）

vessel ['vesl]　　n. 船

viable ['vaiəbl]　　adj. 可行的；能养活的；能生育的

video (conferencing) ['vidiəu]　　n. 电视会议

voluptuous [və'lʌptʃuəs]　　adj. 性感的，体态丰满的

waistband interlining　　腰带衬布

weed out　　清除（不合格的人或物），淘汰

weigh [wei]　　v. 权衡，盘算，仔细考虑

wholesale [həulseil]　　n. 批发

workshop ['wəːkʃɔp]　　n. 工作室；工厂，作坊

woven fabric　　梭织物，机织织物

WPA C With Particular Average　　单独海损赔偿，担保单独海损，水渍险

yardage ['jɑːdidʒ]　　n. 码数

参考文献

[1] 郭平建,况灿,等. 服装英语[M]. 北京:高等教育出版社,2004.

[2] 吕逸华. 服装英语[M]. 2版. 北京:中国纺织出版社,2004.

[3] 郁苓,张玉平,等. 实用英文示范文本[M]. 北京:华语教学出版社,2002.

[4] 廖瑛,莫再树,等. 实用英语应用文写作[M]. 长沙:中南大学出版社,2003.

[5] STEPHENS G. Frings Fashion:from Concept to Consumer[M]. 8th ed. New Jersey:Pearson Education,Inc. 2005.

[6] LEE S. Tate Inside Fashion Design[M]. 5th ed. New Jersey:Pearson Education,Inc. 2004.

[7] SUZANNE G. MARSHALL, HAZEL O. JACKSON, STANLEYMS, et al. Touchie-Specht Individuality in Clothing Selection and Personal Appearance[M]. 5th ed. New Jersey:Prentice Hall,1999.

[8] DIAMOND JAY, DIAMOND ELLEN. The World of Fashion[M]. 3rd ed. Fairchild Publications,Inc. 2002.

[9] Y Li. The Science of Clothing Comfort[J]. Textile Progress,2001,31(1-2):1-135.

[10] REBECCAH PAILES-FRIEDMAN. Smart Textile for Designers (Inventing the Future of Fabrics)[M]. London:Laurence King Publishing,2016.

[11] Power of 3D printing shown at the Met Gala[J]. Colourage,2019,66(5):104-105.

[12] NAYAK, R., CHATTERJEE, K. N., KHURANA, G. K., & KHANDUAL, A. RFID:Tagging the new ERA[J]. Man-Made Textiles in India,2007,50(5):174-177.